Water Pollution

MAN AND THE ENVIRONMENT INFORMATION GUIDE SERIES

Series Editor: Seymour M. Gold, Associate Professor of Environmental Planning, Division of Environmental Planning and Management, University of California, Davis

Also in this series:

ENVIRONMENTAL ECONOMICS—*Edited by Barry C. Field and Cleve E. Willis*

ENVIRONMENTAL EDUCATION—*Edited by William B. Stapp and Mary Dawn Liston*

ENVIRONMENTAL LAW—*Edited by Mortimer D. Schwartz*

ENVIRONMENTAL PLANNING—*Edited by Michael J. Meshenberg*

ENVIRONMENTAL TOXICOLOGY—*Edited by Robert L. Rudd*

ENVIRONMENTAL VALUES, 1860-1972—*Edited by Loren C. Owings*

NOISE POLLUTION—*Edited by Clifford R. Bragdon*

WASTEWATER MANAGEMENT—*Edited by George Tchobanoglous, Robert Smith, and Ronald Crites*

The above series is part of the

GALE INFORMATION GUIDE LIBRARY

The Library consists of a number of separate series of guides covering major areas in the social sciences, humanities, and current affairs.

General Editor: Paul Wasserman, Professor and former Dean, School of Library and Information Services, University of Maryland

Managing Editor: Denise Allard Adzigian, Gale Research Company

Water Pollution

A GUIDE TO INFORMATION SOURCES

Volume 9 in the Man and the Environment Information Guide Series

Allen W. Knight

Professor
University of California, Davis

Mary Ann Simmons

Scientist
Battelle Pacific Northwest Laboratory
Richland, Washington

Gale Research Company
Book Tower, Detroit, Michigan 48226

Library of Congress Cataloging in Publication Data

Knight, Allen W
Water Pollution.

(Man and the environment information guide series ; v. 9) (Gale information guide library)
Includes indexes.
1. Water—Pollution—Documentation. I. Simmons, Mary Ann, joint author. II. Title.
TD423.K54 363.7'394'07 73-17537
ISBN 0-8103-1346-4

VITAE

Allen W. Knight is currently a professor of hydrobiology in the Department of Land, Air and Water Resources at the University of California at Davis. He received his B.S. from Western Michigan University, his M.S. from Michigan State University, and his Ph.D. from the University of Utah.

Knight currently teaches courses titled "The Ecology of Polluted Waters," "The Biology of Running Waters," and "The Biology of Streams." The first two are undergraduate courses while the latter is a graduate course. His research interests include water quality in aquatic systems, the effect of environmental factors on aquatic life, aquaculture, and alternative treatment of wastewater. He has authored or coauthored over eighty papers and reports dealing with various hydrobiological topics.

Mary Ann Simmons is currently employed by Battelle Pacific Northwest Laboratory as a scientist. From 1972 to 1978 she was employed by the University of California as staff research associate, working in the area of hydrobiology and specializing in evaluating water quality and environmental factors as they affect aquatic life.

Simmons received her B.S. in zoology from the University of California at Davis and her M.S. in ecology from the University of California at Davis. She has authored or coauthored over ten articles, papers, and reports in the area of hydrobiology.

CONTENTS

Contents

Contents

FOREWORD

The rapidly expanding scope and sophistication of literature in the area of environmental studies has created the need for an authoritative, annotated guide to the range and quality of information sources. New legislation, publications, developing techniques, and increasing levels of public involvement in environmental issues have created a need for government, industry, organizations, and individuals to seek more authoritative information on man-environment relationships. This information can be used to rationalize proposed public or private actions, resolve issues, and measure the impact of past actions on man and the environment.

The objective of this series is to provide an authoritative and systematic guide to significant information sources on selected topics for use by scholars, students, scientists, reference librarians, consultants, professionals, and citizens. The scope and scale of each volume are directed toward accommodating both the immediate and continuing needs of users in addition to establishing a chronological benchmark of the literature on each topic.

This series is concerned with the cause and effect relationships of man's impact on the urban and natural environment. It emphasizes the human problems, issues, and implications of these relationships that can be used to study or solve environmental problems. The series is intended to be a primary reference for people not familiar with the location and organization of information sources in the general area of environmental studies. It should also serve as a basic reference for advanced students and professionals who need an annotated and selected sample of the current literature.

Each volume is prepared by a noted authority on the topic. The annotations and selection of references have been carefully reviewed to present an objective and balanced array of thought and practice. The organization of each volume is tailored to emphasize the way information is organized in this multidisciplinary and rapidly evolving field.

This volume is an authoritative, current, and comprehensive reference work on water pollution. It brings together a diverse set of information sources from the physical, social, and natural sciences in the economic context of the environmental planning and management process. The scope, historical perspective, and current focus offer an annotated guide to the literature on water pollution.

Foreword

This volume presents a compendium of reference material, case studies, and methodology ideas valuable to the scholar and practitioner. It contains a selected collection of periodical literature, government documents, professional journals, and teaching materials on the rapidly evolving field of water pollution. The introductions, annotations, and sources have been carefully written to give the reader a technical and pragmatic grasp of this fundamental aspect of man-environment relationships.

The scholarly nature of annotations, classification, and cross-referencing of materials presents an integrated view of time, space, values, and events that is not common to other reference works in this area. The authors have made a very technical and diffuse subject accessible to practitioners and students by their clear and concise organization of a wide range of sources.

The authors' treatment of complex and controversial subjects such as: water rights, aquatic ecosystems, aquaculture, recreation management models, environmental impact, and measuring the benefits or costs of policy options is exceptional. This volume focuses on the interface between environmental planning, ecology, land use, and policy analysis, to narrow an important gap in the reference literature.

As the ninth volume in this series, WATER POLLUTION continues the standard of scholarly excellence and problem-solving usefulness followed in other volumes in the Man and the Environment Series. When the entire series is completed in 1980 it will represent the most authoritative and comprehensive guide to the man-environment literature ever published.

Seymour M. Gold
Series Editor

PREFACE

This guide is intended to assist the reader in locating the various aspects of water pollution literature. By virtue of its intended purpose it is neither a complete nor current coverage of all water pollution literature. The authors have consolidated the literature and sources of information into chapters considered most useful to the reader.

The guide was prepared to assist readers with various backgrounds and previous exposure to the field of water pollution. If the reader has had little previous exposure and needs an introduction to water pollution, a selected list of readings is provided in appendix B. This list was prepared in response to numerous requests for a single book which would provide the reader with an overview to the topic of water pollution. This selected reading list has been prepared and tested for nearly a decade. It is a little inconvenient to seek out journal articles and chapters to several books, but no single book adequately provides the reader with the introduction and background contained in the readings. With the selected readings completed, the reader with limited previous experience will be able to evaluate the immense literature associated with water pollution. Those already familiar with water pollution principles can bypass the selected readings and consult the guide according to their needs. The guide is divided into appropriate chapters for ease of use. Sections on abstracts and bibliographic information are provided for those seeking information on a selected topic or specific pollutant. Information services are included for those desiring to obtain literature searches or data retrieval services.

Chapter 4 includes information on the use and availability of guides and bibliographies to state and federal government publications. Chapters 5-9 describe the sources of information contained in dissertations, handbooks, directories, lexicons, encyclopedias, and atlases. Chapter 10 details journals and newsletters, and will be of interest to those seeking current information on aspects of water pollution. The journals included all carry, on a regular basis, water pollution articles. The guide lists journals according to the major subject area covered, thus providing the reader with a guide to those journals most closely related to water pollution topics. This chapter also includes a newsletter section which will be helpful as a means of obtaining current events in a given area, as well as notices of scientific meetings, job opportunities, address changes of scientists, and recent articles of interest.

Part II, "Water Pollution Reference Sources," is of interest because it provides the reader with books and journal articles on various aspects of water pollution. The references are included according to important topic headings. Chapter 12 is included for the reader in need of information on aquatic systems that are unpolluted. Chapter 13, "Methods and Procedures," will be of interest to those either contemplating field evaluations or evaluating impact statements. Chapter 14, on general water pollution references, includes books and articles dealing more or less with a broad portion of the water pollution literature. Chapter 15 is extremely important as it provides the reader with references dealing with specific water pollutants. This section provides a good assemblage of references dealing with specific pollutants, but the reader is encouraged to consult the abstracts and bibliography sources as references in order to obtain current references.

Chapter 16, on pollution sources, will be valuable to the reader seeking information on pollutants from major sources. This may be useful if the reader is not aware of the pollutants from major sources and can consult the section on references of selected pollutants once they are identified by sources.

Wastewater treatment references are included in this guide, in chapter 17, even though other guides in this series provide the reader with greater insight. The topic is extremely important in the area of water pollution as it is the one major means of pollution abatement.

Social, economic, and political aspects references are included in chapter 18, and case studies are in chapter 19. Although they are not references dealing with water pollution per se, the references are important in order to obtain the entire impact of water pollution.

The remaining guides to book reviews, scientific meetings, careers, grant sources, and environmental legislation in part III, chapters 20-26, complete the sources of information included in this guide.

The annotated references under the various topics have generally been limited to those of greatest assistance to the reader in order to keep this volume of a useful size. In most cases journal articles were included without annotations to provide representative coverage. These additional journal articles are included to allow the reader easy access to some journal sources, but are less useful than the annotated references.

A glossary of commonly encountered terms is presented in appendix A. The selected reading list discussed earlier is presented as appendix B.

ACKNOWLEDGMENTS

In the course of preparing this guide we have incurred numerous obligations. We thank Cheryl Stark and Marian Stephenson who assisted in the literature search, and Kim and Pam Knight who assisted in typing and proofing the manuscript.

Special appreciation is due Barbara Knight who typed the manuscript and rendered invaluable editorial assistance.

ABBREVIATIONS

The following abbreviations for organizations, governmental agencies, and terms are frequently encountered in water pollution literature.

ASCA	Automatic Subject Citation Alert
AWWA	American Water Works Association
BOP	Biochemical Oxygen Demand
CEQ	Council on Environmental Quality
EPA	U.S. Environmental Protection Agency
ENDEX	Environmental Data Index
HEW	U.S. Department of Health, Education and Welfare
ISI	Institute for Scientific Information
NAS	National Academy of Sciences
NIH	National Institutes of Health
NMFS	National Marine Fisheries Service
NOAA	National Oceanic and Atmospheric Administration
NSF	National Science Foundation
NWC	National Water Commission
NWF	National Wildlife Federation
OATS	Original Article Tear Sheets
OWRT	Office of Water Research and Technology
SIC	Scientific Information Center
STORET	Storage and Retrieval System--EPA
USDA	U.S. Department of Agriculture
USF & WS	U.S. Fish and Wildlife Service
WPCF	Water Pollution Control Federation
WRC	Water Resources Council, also Water Research Center

Part I

GUIDE TO GENERAL REFERENCE SOURCES

Chapter 1
ABSTRACTS AND INDEXES

While the objective of this bibliography is to provide the reader with an introduction to water pollution literature, it can be neither complete nor current. Access to the vast amount of relatively current literature dealing with water pollution is provided by abstracts and indexes, with most of those in the area of pollution and environmental degradation being of recent origin (1970--). The topic of water pollution encompasses many areas and the reader's scope of interest will dictate which index to consult. Abstracts covering specific topics in water pollution such as EUTROPHICATION and PESTICIDES ABSTRACT will be listed under the appropriate headings in part II of this book.

A. WATER POLLUTION AND GENERAL POLLUTION

ENVIRONMENT ABSTRACTS. New York: Environment Information Center, 1971-- . Monthly with annual cumulative index.

> Prior to 1974 was known as ENVIRONMENT INFORMATION ACCESS, volumes 1-3. Abstracts literature in areas of air pollution, water pollution, and their impact on wildlife. Also includes abstracts pertinent to sections of the FEDERAL REGISTER, reviews films and lists new books and conferences. See ENVIRONMENT INDEX, below.

ENVIRONMENTAL PERIODICALS BIBLIOGRAPHY. Santa Barbara, Calif.: Environmental Studies Institute, 1972-- . Bimonthly with annual cumulative index.

> Publishes table of contents from journals which contain articles on all aspects of environmental degradation.

ENVIRONMENT INDEX. New York: Environment Information Center, 1971-- . Annual.

> Cumulative index for ENVIRONMENT ABSTRACTS, above. Includes, in addition to index, a review of environmental legislation, court decisions and governmental announcements; a synopsis of major federal bills; and lists of conferences, pollution control officials, patents, books, and films. Copies of documents available in hard copy or microfiche.

Abstracts and Indexes

HYDATA: WATER RESOURCES INDEX. Urbana, Ill.: American Water Resources Association, 1965-- . Monthly.

Publishes the table of contents from those journals which contain articles related to water resources ranging from AMERICAN WATER WORKS to LIMNOLOGY AND OCEANOGRAPHY. Includes a section of selected articles of interest from other journals and the table of contents of proceedings and reports.

POLLUTION ABSTRACTS. La Jolla, Calif.: Pollution Abstracts, 1970-- . Semimonthly with annual cumulative index.

Abstracts of articles in the area of air pollution, water pollution, solid wastes, noise, pesticides, radiation, and general environment quality. Includes a calendar of events and a list of special publications and books. Retrieval and literature searches available to subscribers.

SELECTED WATER RESOURCES ABSTRACTS. Washington, D.C.: U.D. Department of Interior, Water Resources Scientific Information Center, 1969-- . Semimonthly with annual cumulative index.

Abstracts of current and earlier monographs, journal articles, conferences, and so forth, in the areas of water chemistry, hydrology, water resources, and water pollution. Also includes grants, facilities, scientific, and technical information.

WATER QUALITY ABSTRACTS. La Jolla, Calif.: Pollution Abstracts, 1973-74. 2 vols.

Summarizes recent literature in the areas of wastewater treatment, industrial wastewater treatment, freshwater pollution, and so forth. Volume 1 covers 1969-72, and volume 2 covers 1973.

WATER QUALITY CONTROL DIGEST. Troy, Mich.: University Digest Services, 1969-- . Bimonthly.

Formerly WATER POLLUTION DIGEST, vol. 1. Abstracts from more than one hundred scientific journals plus government reports, other abstract services, and indexes. Covers all areas of water resources and pollution, plus government programs, laws, and court cases.

WRC INFORMATION. Stevenage, Engl.: Water Research Centre, June 1974-- . Weekly.

Formerly WATER POLLUTION ABSTRACTS and Water Research Association library lists. Abstracts of articles concerning water resources, water quality, monitoring and analysis of water and wastes, water treatment and supply, water use, sewage, industrial effluents, and effects of pollutants.

B. GENERAL INDEX

SCIENCE CITATION INDEX. Philadelphia: Institute for Scientific Information, 1961-- . Quarterly with annual cumulations, also a five-year cumulation 1965-69.

> General index of more than twenty-four hundred scientific journals. Searches based on literature citations obtained from published articles. Since 1968 searches can also be based on key words from the title.

C. SPECIALIZED INDEXES

1. Ecology

APPLIED ECOLOGY ABSTRACTS. London: Information Retrieval, 1975-- . Monthly.

> Abstracts of recent literature in areas of ecology including water quality and management, aquatic resources, and pollution. Includes meetings and symposia.

AQUATIC SCIENCES AND FISHERIES ABSTRACTS. London: Information Retrieval, 1971-- . Monthly with annual cumulative index.

> Abstracts the literature relating to the geophysical environment, biology of aquatic organisms, ecology, oceanology, and limnology. Includes book notices and reviews, as well as translations and notice of meetings. Formed from union of CURRENT BIBLIOGRAPHY FOR AQUATIC SCIENCES AND FISHERIES, 1958-71 and AQUATIC BIOLOGY ABSTRACTS, 1969-71.

ECOLOGICAL ABSTRACTS. Norwich, Engl.: Geo Abstracts, 1974-- . Bimonthly.

> Abstracts of recent literature in marine, freshwater, and terrestrial ecology. Includes biological effects of various pollutants.

OCEANIC ABSTRACTS. La Jolla, Calif.: Oceanic Abstracts, 1964-- . Bimonthly with annual index.

> Abstracts of world literature on the oceans, includes pollution. Also a calendar of meetings. Prior to 1972 known as OCEANIC INDEX.

2. Biology

ALGAE ABSTRACTS. 2 vols. New York: IFI/Plenum Publishing, 1973.

Abstracts covering the relationship between algae and water, and water purification. Compiled from the information base for SELECTED WATER RESOURCES ABSTRACTS. There are few cited references prior to 1960. Volume 1: covers to 1969, and volume 2, 1970-72.

BIOLOGICAL ABSTRACTS. Philadelphia: Bioscience Information Service, 1926-- . Semimonthly with semiannual cumulative indexes.

Abstracts world literature in theoretical and applied biology. Water pollution is included under effects on flora and fauna.

BIOLOGICAL AND AGRICULTURAL INDEX. New York: H.W. Wilson, 1916-- . Monthly, every third issue cumulation of previous two.

Index to English-language periodicals in the areas of agricultural chemicals, biology, botany, ecology, pesticides, and so forth. Useful for articles on water pollution as related to agriculture. Prior to September 1964, known as THE AGRICULTURAL INDEX.

BIORESEARCH INDEX. Philadelphia: Bioscience Information Service, 1965-- . Monthly with annual notices.

Embraces same topics as BIOLOGICAL ABSTRACTS, but indexes symposia, meetings, reviews, letters, notes, theses, government reports, and so forth.

CURRENT CONTENTS: AGRICULTURE, BIOLOGY AND ENVIRONMENTAL SCIENCES. Philadelphia: Institute for Scientific Information, 1970-- . Weekly.

Publishes the tables of contents from journals which contain articles pertinent to the areas specified in the title. Includes a list of the senior authors and their addresses for reprint requests. Each issue contains a list of current books with their tables of contents and book reviews from other journals.

ENVIRONMENTAL HEALTH AND POLLUTION CONTROL. Amsterdam: Excerpta Medica, 1971-- . Monthly with annual subject and author index.

Abstracts of articles relating to the sources of pollutants and effects on man, animals, and the environment; measurement of pollutants, criteria; standards, and the legal, administrative, and socioeconomic aspects of all types of pollutants.

MICROBIOLOGY ABSTRACTS. London: Information Retrieval, 1965-- . Monthly with annual indexes.

Information on water pollution can be found in all three sections: Section A, "Industrial and Applied Microbiology," 1966-- .; Section B, "Bacteriology," 1966-- .; Section C, "Algology, Mycology, and Protozoology," 1972-- . Formerly INDUSTRIAL MICROBIOLOGY ABSTRACTS, 1965-66.

3. Chemistry and Physics

ANALYTICAL ABSTRACTS. London: Society for Analytical Chemistry, 1954-- . Monthly with semiannual cumulations.

Abstracts of articles related to the analysis of compounds including those occurring in water, effleunts and pesticides.

APPLIED SCIENCE AND TECHNOLOGY INDEX. New York: H.W. Wilson, 1958-- . Annual.

Cumulative subject index to English-language periodicals in physical sciences. Water pollution is covered primarily from a physical and chemical aspect.

A.S.C.E. PUBLICATIONS ABSTRACTS. New York: American Society of Civil Engineers, 1966-- . Bimonthly.

Abstracts of all papers published in A.S.C.E. journals including JOURNAL OF ENVIRONMENTAL ENGINEERING DIVISION and CIVIL ENGINEERING. Primarily covers water pollution as related to water treatment.

CHEMICAL ABSTRACTS. Columbus, Ohio: American Chemical Society, 1907-- . Biweekly since 1967. 10-year indexes to 1957, 5-year indexes through 1972, and semiannual indexes since 1972.

Abstracts literature of chemistry and allied fields, covers water pollution from a chemical and chemical engineering aspect.

NUCLEAR SCIENCE ABSTRACTS. Washington, D.C.: U.S. Atomic Energy Commission, 1948-- . Seminonthly with semiannual cumulative indexes.

Covers scientific and technical reports which relate to radioactive compounds. Includes radioactive compounds in the environment and its effect on organisms.

Chapter 2
BIBLIOGRAPHIES

The following bibliographies are those dealing with the general topics of the environment and water pollution. Specific bibliographies on such topics as mercury, oil, and eutrophication will be found under those headings. Although bibliographies provide the same information as indexes, those references listed in the previous section are of a continuing nature, while those presented here are not. Thus, most of these references will be somewhat dated, but they do provide needed background information and a starting point for literature searches.

A. WATER POLLUTION

Alexander, Helen, ed. EFFECTS OF POLLUTION ON FISH. Washington, D.C.: National Agricultural Library, 1970. 11 p.

> Bibliographic citations of articles published between 1950 and 1969 which deal with the effects of pollution on fish.

ESTUARINE POLLUTION: A BIBLIOGRAPHY. Vol. 3. Washington, D.C.: Water Resources Scientific Information Center, 1976. Distributed by National Technical Information Service, Springfield, Va. 566 p.

> Abstracts on general estuarine pollution taken from SELECTED WATER RESOURCES ABSTRACTS from April 1974 to September 1976. Volume 1 covered the period October 1968 through December 1972; volume 2, January 1973 to April 1974.

Sinha, Evelyn. COASTAL/ESTUARINE POLLUTION: AN ANNOTATED BIBLIOGRAPHY. Ocean Engineering Information Series, vol. 3. La Jolla, Calif.: Ocean Engineering Information Service, 1970. x, 87 p.

> An annotated bibliography of 631 publications published between 1965 and May 1970 on various aspects of hydrology and pollution of coastal waters.

_____. LAKE AND RIVER POLLUTION. Ocean Engineering Information Series, vol. 4. La Jolla, Calif.: Ocean Engineering Information Service, 1971. ix, 85 p.

An annotated bibliography covering the period 1968-70 inclusive, comprising such areas as hydrological processes, effects on biota, and parameters of pollutants.

SUBSURFACE WATER POLLUTION: A SELECTED ANNOTATED BIBLIOGRAPHY. 3 vols. Washington, D.C.: Water Resources Scientific Information Center, 1972.

References taken from SELECTED WATER RESOURCES ABSTRACTS over the period October 1968 through December 1971; covers subsurface waste injection; saline water intrusion, and percolation from surface sources.

WATER POLLUTION. Report DDC-TAS, 73-26. Alexandria, Va.: Defense Documentation Center, 1973. Distributed by National Technical Information Service, Springfield, Va. 298 p.

Contains over two hundred citations covering the period 1953 to February 1973. Concerned with causes, detection, control, treatment, and elimination of water pollution and pollutants.

B. ENVIRONMENT

Bennett, G.F., and Bennett, J.C. ENVIRONMENTAL LITERATURE: A BIBLIOGRAPHY. 3d ed. Park Ridge, N.J.: Noyes Data, 1973. viii, 134 p.

Bibliographic citations of the environmental literature. Good section on water. Includes books, pamphlets, journals, and audiovisual aids, but no journal articles.

Dasbach, Joseph M. SCIENCE FOR SOCIETY: A BIBLIOGRAPHY. 6th ed. Washington, D.C.: American Association for the Advancement of Science, 1976. x, 104 p.

Bibliographic citations of publications pertaining to various aspects of the environment including water resources and pollution.

ENVIRONMENT: AN ANNOTATED BIBLIOGRAPHY. Rome: United Nations, Food and Agricultural Organization, 1971. 782 p.

Index to publications and documents issued by the Food and Agricultural Organization between 1967-70. Includes articles on water pollution and pollutants. Copies of indexed articles available from Food and Agricultural Organization.

Hammond, K.A.; Macinko, G.; and Fairchild, W.B., eds. SOURCEBOOK ON THE ENVIRONMENT: A GUIDE TO THE LITERATURE. Chicago: University of Chicago Press, 1978. x, 613 p.

> A broad guide to selected literature readily available through college libraries, the SOURCEBOOK includes chapters on "Water and Water Quality" and "Impact of Industrial Activity on Water Quality."

Minneapolis Public Library. ECOL: A BOOK CATALOG. Chicago: American Library Association, 1974. vii, 201 p.

> Primarily an author and subject index of books from all subject areas which relate to man's physical environment and his impact on it. This index contains general information as well as information relating specifically to Minnesota.

Watkins, J.B., ed. ECOLOGY AND ENVIRONMENTAL QUALITY: A SELECTED AND ANNOTATED BIBLIOGRAPHY FOR BIOLOGISTS AND EARTH SCIENTISTS. Syracuse, N.Y.: Syracuse University Libraries, 1971. 127 p.

> Selected bibliography of information on the biological aspects of air, water, and land pollution available from Syracuse University.

Winton, H.N.M., ed. MAN AND THE ENVIRONMENT: A BIBLIOGRAPHY OF SELECTED PUBLICATIONS OF THE UNITED NATIONS SYSTEM, 1946-1971. New York: Unipub, 1972. xxi, 305 p.

> Annotated bibliography of publications covering a broad range of topics including water resources and water pollution, issued by the United Nations. Contains filmstrips, directories, and other bibliographies.

Chapter 3
INFORMATION SERVICES

With the growth in the volume of literature being published, searching through abstracts and indexes can be very time consuming. Also, some abstracts or the article of interest may be unavailable. There are numerous private and governmental agencies which provide (usually for a fee) manual and computerized literature searches, as well as interlibrary loans or copying services. Further, many university and large public libraries now possess the capacities for providing literature searches. The following references will include, first, those groups which provide literature searches, followed by libraries (listed by state) which have specialized holdings related to water and the environment. All of these are open to the public, at least for reference use. Other services which may be offered by these groups include translations, compilation of bibliographies, state-of-the-art reviews (generally on a contract basis), and consultation.

The last category contains governmental agencies which provide data retrieval systems. Data is compiled from federal, state and private sources and is generally available to nongovernmental organizations and individuals.

Additional information on these references can be obtained from:

Kruzas, Anthony T., ed. ENCYCLOPEDIA OF INFORMATION SYSTEMS AND SERVICES. 3d ed., rev. Detroit: Gale Research Co., 1978. xiii, 1,030 p.

Young, Margaret Labash, and Young, Harold Chester. DIRECTORY OF SPECIAL LIBRARIES AND INFORMATION CENTERS. 5th ed. 3 vols. Detroit: Gale Research Co., 1979.

______. SUBJECT DIRECTORY OF SPECIAL LIBRARIES AND INFORMATION CENTERS. Vol. 5: SCIENCE AND TECHNOLOGY LIBRARIES. 5th ed. Detroit: Gale Research Co., 1979. xx, 277, 85 p.

A. LITERATURE SEARCHES—PRIVATE

BIOSIS, Bioscience Information Service of Biological Abstracts, 2100 Arch Street, Philadelphia, Pa. 19103.

BIOSIS covers all areas of the biological sciences. Its publications include BIOLOGICAL ABSTRACTS and ABSTRACTS OF HEALTH EFFECTS OF ENVIRONMENTAL POLLUTANTS. Search services offered include Bioresearch Today, which is a series of monthly journals containing abstracts on a particular subject area such as pesticides. Personalized computer literature searches are available through C.L.A.S.S. (Current Literature Alerting Search Service) and a retrospective search service.

Center for Short-Lived Phenomena, Smithsonian Institution, Office of Environmental Science, 60 Garden Street, Cambridge, Mass. 02138.

Center's function is to collect and quickly disseminate information on shortlived phenomena, including major pollution occurrences. Center provides a reference and referral service, as well as publishing Event Notification Report cards, annual reports, and reviews.

Chemical Abstracts Service, American Chemical Society, Ohio State University, Columbus, Ohio 43210

Covers all aspects of the chemical and chemical engineering literature published in CHEMICAL ABSTRACTS. Literature searches are provided by CA CONDENSATES and CA SUBJECT INDEX ALERT. Microreproduction is also available.

Environment Information Center, Ecology Forum, 124 East 39th Street, New York, N.Y. 10016.

Center is concerned with all areas of the environment, particularly conservation and pollution. It publishes ENVIRONMENT ABSTRACTS, ENVIRONMENT INDEX, and ENVIRONMENT REGULATIONS HANDBOOK. Offers special literature searches and ENVIROFICHE, copies of abstracted items in microfiche.

ISI (Institute for Scientific Information), 325 Chestnut Street, Philadelphia, Pa. 19106.

Journal literature in the sciences and technical fields are covered in the following publications: SCIENCE CITATION INDEX, CURRENT CONTENTS, and SOCIAL SCIENCE CITATION INDEX. Services incorporate generalized searches of selected topics including water pollution (ASCA TOPICS), personalized searches (ASCA), and retrospective searches (ISI Search Service). Original journal articles are also available through OATS.

Macmillan Information, 866 Third Street, New York, N.Y. 10022.

Wide range of literature topics are available through a number of taped data bases. These include TRANSDEX--a compilation of translated books and articles; PANDEX--citations to the international scientific and technical literature; MARC II-- a record of books cataloged by the U.S. Library of Congress; plus tapes used for the BIBLIOGRAPHY OF AGRICULTURE AND ALLIED SCIENCES, and GOVERNMENT REPORTS ANNOUNCEMENTS.

Pollution Abstracts, P.O. Box 2369, La Jolla, Calif. 92037.

Covers the pollution literature and that concerning the oceans in POLLUTION ABSTRACTS and OCEANIC ABSTRACTS, respectively. Offers manual literature searches on a contract basis and document retrieval for subscribers.

Recon, Water Information Section, University of Arizona, Tucson, Ariz. 85721.

Computerized information retrieval system whose information bases include WATER RESOURCES ABSTRACTS, plus data bases on toxic materials, mercury, and heated effluents.

Smithsonian Science Information Exchange, Smithsonian Institution, 1730 M Street, N.W., Room 300, Washington, D.C. 20036

Exchange is the central clearinghouse for ongoing research sponsored by both federal and nonfederal organizations. Indicates who's working on what, where, and with what support.

Unipub, Box 433, Murray Hill Station, New York, N.Y. 10016.

Unipub is the U.S. distributor of United Nations publications. It issues a checklist of publications on various topics including water. United Nations publications may be ordered from Unipub.

Water Research Centre, Stevenage Laboratory, Elder Way, Stevenage, Herts SG1 1TH, England.

Computerized information retrieval system whose information base is material published in WRC INFORMATION and supplementary abstracts including individual conference papers where proceedings have been published. Translations of some papers. abstracted in WRC INFORMATION are available and quotations are given for cost of translating any paper.

Xerox University Microfilms, 300 North Zeeb Road, Ann Arbor, Mich. 48106.

Offers computerized searches of the doctoral dissertation literature.

B. LITERATURE SEARCHES—FEDERAL GOVERNMENT

Ecological Sciences Information Center, U.S. Atomic Energy Commission, Oak Ridge National Laboratory, P.O. Box X, Building 2029, Oak Ridge, Tenn. 37830.

> Center compiles literature related primarily to radioecology and thermal effects. It publishes specialized bibliographies and offers literature searches.

Environmental Information System Office, U.S. Atomic Energy Commission, Oak Ridge National Laboratory, P.O. Box X, Oak Ridge, Tenn. 37830.

> Office coordinates information activities of environmental research projects at the laboratory. It maintains an environmental directory which is a compilation of information centers, research projects, and individuals, and it publishes TRACE CONTAMINANTS ABSTRACTS. Services offered include literature searches, microreproduction, copying, and interlibrary loan, all of which are available to U.S. government agencies, their contractors, and research and educational institutions, as well as industry. Two of its component information centers are the Toxic Material Information Center, which compiles literature on metals and organometallic compounds, and the Toxicology Information Response Center, which is the national and international center of toxicology information. Services offered by these two centers are generally available without restriction.

Information Center and Library, U.S. Environmental Protection Agency, National Environmental Research Center, Cincinnati, Ohio 45268.

> Primary compiler of scientific and technical information for EPA. Offers literature searches and interlibrary loans; the former is available only to EPA staff, consultants, and contractors, while the latter is available to cooperating libraries.

NTIS (National Technical Information Service), U.S. Department of Commerce, 5285 Port Royal Road, Springfield, Va. 22151.

> Central source for the sale of government sponsored research reports prepared by federal agencies, their contractors, or grantees. It compiles general subject bibliographies and performs personalized searches (NTISearch). It also publishes WEEKLY GOVERNMENTAL ABSTRACTS and abstracts on more selected topics such as ENVIRONMENTAL POLLUTION and CONTROL, as well as GOVERNMENT REPORTS ANNOUNCEMENTS and GOVERNMENT REPORTS INDEX.

OASIS (Oceanic and Atmospheric Science Information system), U.S. Environmental Data Service/National Oceanic and Atmospheric Administration, 3300 Whitehaven Street, N.W., Washington, D.C. 20235.

System, along with ENDEX (Environmental Data Index), was designed to be a "one-stop environmental data and information referral service." OASIS is an information retrieval system which references the published technical literature and ongoing research efforts in the environmental sciences and marine and coastal resources. OASIS has access to both NOAA and non-NOAA bibliographic data bases. These include the data bases of Biosis; Chemical Abstract Service; ISI; National Oceanographic Data Center (Biological Information Retrieval System); Congressional Information Service, Inc.; Denver Public Library (Fish and Wildlife Reference Service); NTIS (GOVERNMENT REPORTS ANNOUNCEMENT); Cornell University (HEATED EFFLUENTS); Energy Research and Development Administration (NUCLEAR SCIENCE ABSTRACTS); Data Courier, Inc. (OCEANIC SCIENCE ABSTRACTS, POLLUTION); U.S. Department of Interior (SELECTED WATER RESOURCES ABSTRACTS); Oak Ridge National Laboratory (TOXIC MATERIALS); and Smithsonian Science Information Exchange.

Water Resources Scientific Information Center, U.S. Department of the Interior, Office of Water Resources Research, 18th and C Street, N.W., Washington, D.C. 20240.

Covers all areas pertaining to water; publishes SELECTED WATER RESOURCES ABSTRACTS, as well as special subject bibliographies on such topics as MERCURY IN WATER, DETERGENTS IN WATER, and so forth. Personalized computer searches are also available.

C. LIBRARIES

The following list of libraries are those institutions which generally restrict their holdings to a specialized area, in this case some aspect of water pollution. Only those libraries open to the public were included, however, there are usually some restrictions on usage of the facilities.

ALABAMA

International Paper Company
Erling Riis Research Laboratory
Technical Library
P.O. Box 2328
Mobile, Ala. 36601

ALASKA

U.S. National Marine Fisheries Service
Auke Bay Fisheries Laboratory
Fisheries Research Library
P.O. Box 155
Auke Bay, Alaska 99821

ARKANSAS

Arkansas Department of Pollution Control and Ecology
Technical Information Center
8001 National Drive, Box 9583
Little Rock, Ark. 72209

CALIFORNIA

Aquatic Research Institution Library
2193 Curtis Street
Hayward, Calif. 94545.

Lake Tahoe Area Council Library
P.O. Box 3475
South Lake Tahoe, Calif. 95705

Oil Spill Information Center
Office of Librarian
University of California
Santa Barbara, Calif. 93106

U.S. Environmental Protection Agency
Region 9 Library
100 California Street
San Francisco, Calif. 94111

University of California, Berkeley
Water Resources Center Archives
North Gate Hall, Room 40
Berkeley, Calif. 94720

University of California
Scripps Institution of Oceanography
La Jolla Library
La Jolla, Calif. 92037

COLORADO

Denver Public Library
Conservation Library Center
1357 Broadway
Denver, Colo. 80203

DISTRICT OF COLUMBIA

U.S. Environmental Protection Agency
Headquarters Library
401 M Street, S.W.
Washington, D.C. 20460

Water Pollution Control Federation Library
3900 Wisconsin Avenue
Washington, D.C. 20016

ILLINOIS

Illinois Institute for Environmental Quality Library
390 West Washington
Chicago, Ill. 60606

MAINE

Maine Department of Environmental Protection Library
State House
Augusta, Maine 04330

MICHIGAN

U.S. Bureau of Sport Fisheries and Wildlife
Great Lakes Fishery Laboratory
John Van Oosten Great Lakes Fishery Research Library
1451 Green Road
Ann Arbor, Mich. 48107

University of Michigan
Great Lakes Research Division Library
1077 North University Building
Ann Arbor, Mich. 48104

MINNESOTA

Economics Laboratory, Inc.
Information Center
Osborn Building
St. Paul, Minn. 55102

U.S. Environmental Protection Agency
National Water Quality Laboratory Library
6201 Congdon Boulevard
Duluth, Minn. 55804

MISSOURI

Missouri Clean Water Commission Library
1014 Madison Street
Jefferson City, Mo. 65101

U.S. Bureau of Sport Fisheries and Wildlife
Fish-Pesticide Research Laboratory
Route 1
Columbia, Mo. 65201

Water and Wastewater Technical School
George Ferrel Memorial Library
P.O. Box 370
Neosho, Mo. 64850

NEW HAMPSHIRE

New Hampshire Water Supply and Pollution Control Commission Library
105 Loudon Road
Concord, N.H. 03301

NEW JERSEY

U.S. Environmental Protection Agency
Region 2
Field Office, Technical Library
Edison, N.J. 08817

OKLAHOMA

U.S. Environmental Protection Agency
Robert S. Kerr Environmental Research Laboratory Library
P.O. Box 1198
Ada, Okla. 74820

OREGON

Oregon State University
Water Resources Research Institute and Air Resources Center Library
Covell Hall 115
Corvallis, Oreg. 97331

PENNSYLVANIA

Carnegie-Mellon University
Mellon Institute Library
4400 Fifth Avenue
Pittsburgh, Pa. 15213

Franklin Institute Research Laboratories
Science Information Services
20th and Benjamin Franklin Parkway
Philadelphia, Pa. 19103

Pennsylvania Department of Environmental Resources
Bureau of Water Quality Management
Technical Reference File
Fulton National Building
13th Floor
Harrisburg, Pa. 17101

RHODE ISLAND

University of Rhode Island
New England Marine Resources Information Program
Narragansett Bay Campus
Narragansett, R.I. 02882

TENNESSEE

University of Tennessee
Water Resources Research Center Library
100 White Avenue
Knoxville, Tenn. 37916

WASHINGTON

U.S. Environmental Protection Agency
Region 10 Library
1200 Sixth Avenue
Seattle, Wash. 98101

WISCONSIN

Badger Meter Library
4545 West Brown Deer Road
Milwaukee, Wis. 53223

D. DATA RETRIEVAL

U.S. Council on Environmental Quality. FEDERAL ENVIRONMENTAL MONITORING DIRECTORY. Washington, D.C.: 1973. Distributed by U.S. Government Printing Office. 105 p.

A list of primarily federal agencies where data on various environmental parameters, including water pollutants, can be obtained.

U.S. Environmental Protection Agency, Management Information Data Systems Division, Environmental Information Retrieval On-line (ENVIRON); 401 M Street, S.W., Washington, D.C. 20460.

Data base includes proposed and active research projects, water quality surveillance network data, EPA research project reporting system, and more. Use of ENVIRON is available to non-EPA agencies and organizations on a contract basis.

U.S. Environmental Protection Agency, Office of Air and Water Programs, Monitoring and Data Support Division, Information Access and User Assistance Branch, 401 M Street, S.W., Washington, D.C. 20460.

Storage and retrieval system named STORET, which collects data related to the discharge of pollutants into waterways. System can be accessed throughout the country and is limited to pollution control authorities at all levels of government.

U.S. Geological Survey, Water Resources Division, National Water Data System, 18th and F Street, N.W., Washington, D.C. 20242.

This division coordinates and publishes statistics on surface water supply and quality. Summary reports and data are available to planners, developers, and managers.

U.S. Geological Survey, Water Resources Division, Office of Water Data Coordination, 12201 Sunrise Valley Drive, Reston, Va. 22070

Division coordinates acquisition of water data by federal agencies and publishes a catalog on water data from both federal and nonfederal sources--CATALOG OF INFORMATION ON WATER DATA. Data includes surface and ground water quality and quantity.

U.S. National Oceanic and Atmospheric Administration, Environmental Data Service, National Oceanographic Data Center, Rockville, Md. 20852.

Storage and retrieval of oceanographic data.

U.S. National Oceanic and Atmospheric Administration, Environmental Data Service, National Oceanographic Data Center, ENDEX, 2001 Wisconsin Avenue, N.W., Washington, D.C. 20235.

Data retrieval system associated with OASIS. Its files cover primarily the Great Lakes and eastern coastal areas with expansion planned for the other coastal areas. Information which can be obtained from ENDEX includes the types of parameters and volumes of data available, the methods used to measure them, when and where the data was collected, instrumentation, restrictions on data availability, publications containing the data, whom to contact for information, and the estimated cost of obtaining the data.

U.S. National Oceanic and Atmospheric Administration, National Ocean Survey, Lake Survey Center, 630 Federal Building and U.S. Courthouse, Detroit, Mich. 48226.

Publishes data summaries from surveys of the Great Lakes and their outflow rivers, plus several additional surrounding bodies of water.

Chapter 4
GOVERNMENT DOCUMENTS

The following references are guides to documents published by both federal and state governments. Government publications which relate to specific areas of water pollution, such as PESTICIDES ABSTRACTS, are listed under the appropriate headings.

A. GUIDES TO FEDERAL GOVERNMENT PUBLICATIONS

Andriot, John L. GUIDE TO U.S. GOVERNMENT PUBLICATIONS. 2 vols. McLean, Va.: Documents Index, 1978.

> Annotated guide to current series and periodicals published by government agencies, and complete listing of U.S. Superintendent of Documents issuances since 1900. Volume 1 covers SuDoc classes A-E; volume 2, classes F-O. Volume 3, to be published, will cover classes P-Z. Appendix in each volume lists new SuDoc publications not yet incorporated into individual sections of Guide.

Body, Alexander C. ANNOTATED BIBLIOGRAPHY OF BILBIOGRAPHIES ON SELECTED GOVERNMENT PUBLICATIONS. 4 vols. Kalamazoo: Western Michigan University, 1967, Supplements 1968, 1970, 1972.

> Bibliographic information on bibliographies of government publications plus abstracts. A broad scope of bibliographies is included.

Buchanan, William W., and Kanely, Edna A., eds. CUMULATIVE SUBJECT INDEX TO THE MONTHLY CATALOG OF U.S. GOVERNMENT PUBLICATIONS 1900-1971. 14 vols. Washington, D.C.: Carrollton Press, 1972-74.

MONTHLY CATALOG OF U.S. GOVERNMENT PUBLICATIONS. Washington, D.C.: Superintendent of Documents, U.S. Government Printing Office, 1895-- . Monthly with annual index.

Index of all government publications arranged according to issuing agency. Ordering information included.

SELECTED LIST OF U.S. GOVERNMENT PUBLICATIONS. Washington, D.C.: Superintendent of Documents, U.S. Government Printing Office, 1928-- . Biweekly.

Contains title, price, and annotation of those government publications which have public appeal. Formerly WEEKLY LIST, 1928-51.

B. GUIDES TO FEDERAL GOVERNMENT SPONSORED RESEARCH REPORTS

BIBLIOGRAPHY OF R & D RESEARCH REPORTS. Washington, D.C.: Office of Research and Development, U.S. Environmental Protection Agency, 1972-- . Irregular.

Bibliographic information for government-sponsored research reports on environmental topics, includes the WATER POLLUTION CONTROL RESEARCH SERIES and BIBLIOGRAPHY OF WATER QUALITY RESEARCH REPORTS. Formerly BIBLIOGRAPHY OF R & M RESEARCH REPORTS, July 1972-June 1973.

E.P.A. REPORTS BIBLIOGRAPHY. QUARTERLY SUPPLEMENT. Washington, D.C.: U.S. Environmental Protection Agency, Jan.-Mar. 1977-- .

Updates the EPA CUMULATIVE BIBLIOGRAPHY, 1970-1976. Abstracts plus author, corporate source, subject, contract, and title indexes for all EPA and its predecessor agencies. Reports available from the National Technical Information Service.

GOVERNMENT REPORTS ANNOUNCEMENTS. Springfield, Va.: National Technical Information Service, 1946-- . Biweekly.

Abstracts of all government sponsored research. Citations include research subject, investigator, contractor, and report numbers. Formerly BIBLIOGRAPHY OF SCIENTIFIC AND INDUSTRIAL REPORTS, 1946-49; BIBLIOGRAPHY OF TECHNICAL REPORTS, 1949-54; U.S. GOVERNMENT RESEARCH REPORTS, 1954-64; U.S. GOVERNMENT RESEARCH & DEVELOPMENT REPORTS, 1965-76. Absorbed TECHNICAL TRANSLATIONS.

GOVERNMENT REPORTS INDEX. Springfield, Va.: National Technical Information Service, 1965-- . Biweekly.

Indexes GOVERNMENT REPORTS ANNOUNCEMENTS by subject, personal and corporate authors, and by government contract and order numbers. Formerly U.S. GOVERNMENT RESEARCH AND DEVELOPMENT INDEX, 1965-70.

INDEXED BIBLIOGRAPHY OF OFFICE OF RESEARCH AND DEVELOPMENT REPORTS. Washington, D.C.: Office of Research and Development, U.S. Environmental Protection Agency, 1975. viii, 293 p.

Listing of EPA'S Office of Research and Development final reports on inhouse, grant, contract, and interagency projects for period 1967 through 1974.

NEW R AND D REPORTS. Washington, D.C.: Office of Research and Development, U.S. Environmental Protection Agency, 1972-- . Irregular.

Newsletter type listing of recent research, grant and inhouse reports which were sponsored by EPA. Title, author, report, and project numbers given plus price and source for purchasing. Formerly NEW R AND M REPORTS, 1972-73.

ORD PUBLICATIONS SUMMARY. Washington, D.C.: Office of Research and Development, U.S. Environmental Protection Agency, distributed by National Technical Information Service, 1975-- . Quarterly.

Listing and abstract of recent final reports issued by the Office of Research and Development. This publication is a continuation of INDEXED BIBLIOGRAPHY OF OFFICE OF RESEARCH AND DEVELOPMENT REPORTS (see above) and contains all final reports issued in the previous eighteen months with abstracts only on the most recent.

WATER RESOURCES RESEARCH CATALOG. Washington, D.C.: Office of Water Resources Research, U.S. Department of Interior, 1965-- . Annually.

Bibliographic information on water resources research projects, both federally and nonfederally funded. Includes abstracts.

WEEKLY GOVERNMENT ABSTRACTS: ENVIRONMENTAL POLLUTION AND CONTROL. Springfield, Va.: National Technical Information Service, 1970-- . Weekly.

Contains same information as GOVERNMENT REPORTS ANNOUNCEMENTS (see p. 24), restricted to research related to environmental pollution. Succeeded GOVERNMENT REPORTS TOPICAL ANNOUNCEMENTS, 1970-73.

C. BIBLIOGRAPHIES OF SELECTED GOVERNMENT PUBLICATIONS

Indiana University. Government Publications Department. MAN AND HIS ENVIRONMENT: SELECTED GOVERNMENT PUBLICATIONS 1950-1970. Bloomington: 1970. 95 p.

Subjects covered include environmental pollution, human ecology, and conservation of natural resources. Publications were selected from those issued by federal, state, interstate, city, county, regional, and international government organizations.

Kiraldi, Louis, and Burk, Janet L., eds. POLLUTION: A SELECTED BIBLIOGRAPHY OF U.S. GOVERNMENT PUBLICATIONS ON AIR, WATER AND LAND POLLUTION. Kalamazoo: Western Michigan University, Institute of Public Affairs, 1971. 78 p.

> Bibliographic citations of approximately eight hundred government publications published between 1965-70.

D. GUIDES TO STATE GOVERNMENT PUBLICATIONS

Giefer, G.J., and Todd, D.K., eds. WATER PUBLICATIONS OF STATE AGENCIES. Port Washington, N.Y.: Water Information Center, 1972. 319 p.

> Bibliography of publications on water resources and their management by the states. Includes addresses for obtaining publications and a list of Water Resources Research Centers.

MONTHLY CHECKLIST OF STATE PUBLICATIONS. Washington, D.C.: Library of Congress, 1910-- . U.S. Government Printing Office. Monthly with annual index.

> Bibliographic information on state documents received by the Library of Congress. Information is listed according to state and agency. No subject or author indexes except annually.

Many states issue publication lists which can be useful in locating and obtaining reports of state agencies and commissions. These lists vary considerably as to completeness and regularity of issuance. Some are continuing records of all state documents, while others indicate the reports of a particular agency. Inquiries about publication lists should be directed to the state library.

Chapter 5

DISSERTATIONS, CONFERENCE PROCEEDINGS, AND TRANSLATIONS

A. DISSERTATIONS

AMERICAN DOCTORAL DISSERTATIONS. Ann Arbor, Mich.: Xerox University Microfilms, 1955-- .

> Complete listing of all doctoral dissertations accepted by American and Canadian universities. Bibliographic information only. Formerly INDEX TO AMERICAN DOCTORAL DISSERTATIONS 1955-65.

BIBLIOGRAPHY OF DOCTORAL RESEARCH ON ECOLOGY AND THE ENVIRONMENT. Ann Arbor, Mich.: Xerox University Microfilms, n.d. Pamphlet.

> List of more than nine hundred doctoral dissertations from American universities between 1938 and 1970. Source is DISSERTATION ABSTRACTS INTERNATIONAL.

COMPREHENSIVE DISSERTATION INDEX 1861-1972. 37 vols. Ann Arbor, Mich.: Xerox University Microfilms, 1973.

> Bibliographic citations of doctoral dissertations awarded in the United States between 1861 and 1972. Each volume represents a separate field. Supplement issued in 1974 for 1973 dissertations.

DISSERTATION ABSTRACTS INTERNATIONAL. Section B. Sciences and Engineering. Ann Arbor, Mich.: University Microfilms International, 1938-- . Monthly.

> Includes doctoral dissertations from over three hundred institutions in the United States, Canada, and Europe. Citation includes author, title, university, supervisor, abstract, number of pages, and order number. Dissertations available from publisher. Prior to 1970 called DISSERTATION ABSTRACTS.

B. CONFERENCE PROCEEDINGS

CURRENT PROGRAMS. Chestnut Hill, Mass.: World Meeting Information Center, 1973-- . Quarterly with annual cumulation.

Provides the following information on meetings in the life sciences, chemistry, physical sciences, geoscience, and engineering: title of meeting, date, location, name of sponsoring groups(s), ordering information for publications, and list of papers (name and address of author, complete title, any ordering number).

DIRECTORY OF PUBLISHED PROCEEDINGS. Harrison, N.Y.: InterDok, 1965-- . Monthly (September-June).

Bibliographic directory of conference proceedings and preprints (includes meetings, seminars, and summer schools). Citation includes date, location, title of meeting and principal sponsors, publisher or distributor and his address, order number, and price. InterDok provides an acquisition service for subscribers.

INDEX OF CONFERENCE PROCEEDINGS RECEIVED. Wetherby, U.K.: British Library Lending Division, 1965-- . Monthly.

Proceedings indexed by keywords from the title of the conference proceedings, plus date, location, and acquisition code. Photocopies are available from the publisher.

PROCEEDINGS IN PRINT. Mattapan, Mass.: Proceedings in Print, 1964-- . Bimonthly with annual index.

Index to conference proceedings (includes reports, symposia, lecture series, and courses) in all subject areas and all languages. Provides title of conference, location, data, sponsors, proceeding title, editors, publisher, and date of publication.

C. TRANSLATIONS

GOVERNMENT REPORTS ANNOUNCEMENTS. (See p. 24)

Absorbed TECHNICAL TRANSLATIONS.

TRANSLATIONS REGISTER INDEX. Chicago: National Translations Center, 1967-- . Monthly.

Index of translations from world's literature into English in the natural, physical, medical, and social sciences. Citation includes author, title, where article was originally published, and where it is available.

WORLD INDEX OF SCIENTIFIC TRANSLATIONS AND LIST OF TRANSLATIONS NOTIFIED TO E.T.C. Delft, Netherlands: European Translation Center, 1967-- . Monthly with annual cumulation.

List of translations from nonwestern to western languages (not necessarily English) of serial articles, patents, and standards.

Chapter 6
HANDBOOKS

The following references are useful as sources of statistics on water quality and toxicity. Included is a guideline for preparation of Environmental Impact Statements.

Altman, P.L., and Dittmer, D.S., eds. BIOLOGY DATA BOOK. 2d ed. 3 vols. Bethesda, Md.: Federation of American Societies for Experimental Biology, 1972-74.

> Volume 2 is probably the most useful, as it contains data on temperature tolerances for algae and aquatic organisms.

______. ENVIRONMENTAL BIOLOGY. Biological Handbook Series. Bethesda, Md.: Federation of American Societies for Experimental Biology, 1966. xxi, 694 p.

> Contains data on the hydrologic cycle, physical properties of water, and water turnover in mammals and water in plants. Contains a section on temperature tolerances of aquatic and terrestrial organisms.

Bond, R.G., and Straub, G.P., eds. CRC HANDBOOK OF ENVIRONMENTAL CONTROL. Vol. 3: WATER SUPPLY AND TREATMENT. Cleveland: CRC Press, 1973. 835 p.

> Encompasses many areas of water quality including effect of compounds found in water on human and aquatic life, water standards and analytical methods, water supply, and requirements for agriculture, industry, and public consumption.

FISH KILLS. Washington, D.C.: U.S. Environmental Protection Agency, 1960-- . Annual.

> Tables of statistics indicating where, when, and probable causes of fish kills in the United States. Includes some analysis of the kills.

Handbooks

GUIDELINES FOR PREPARATION OF ENVIRONMENTAL STATEMENTS FOR REVIEWING AND COMMENTING ON ENVIRONMENTAL STATEMENTS PREPARED BY OTHER FEDERAL AGENCIES. Rev. ed. Washington, D.C.: U.S. Environmental Protection Agency, Region X, 1973. Var. pag.

> Cover title reads ENVIRONMENTAL IMPACT STATEMENT GUIDELINES. Presents the National Environmental Policy Act plus other amendments and guidelines pertinent to impact statements. Indicates what impact statements should contain and gives guidelines for specific projects.

Lund, H.F., ed. INDUSTRIAL POLLUTION CONTROL HANDBOOK. New York: McGraw-Hill, 1971. 887 p. Glossary.

> Includes history of industrial pollution plus legislation, standards, control programs, and research. Another section covers pollution from specific types of industries and final section deals with pollution control equipment.

STATISTICAL ABSTRACT OF THE UNITED STATES. Washington, D.C.: U.S. Bureau of the Census, 1878-- . Annual. Distributed by U.S. Government Printing Office.

> Statistics on population, geography, and environment (water use and pollution). Includes federal funds for pollution control, polluting spills, and abatement.

Sunshine, Irving, ed. CRC HANDBOOK OF ANALYTICAL TOXICOLOGY. Cleveland: CRC Press, 1969. xiv, 1,081 p.

> Sections applicable to water pollution include: water analysis and criteria, pesticides in water, and the effects of pesticides on humans and aquatic organisms.

Thomas, William A.; Goldstein, G.; and Wilcox, W.H. BIOLOGICAL INDICATORS OF ENVIRONMENTAL QUALITY. Ann Arbor, Mich.: Ann Arbor Science Publishers, 1973. 254 p.

> Compilation of abstracts from the literature through 1971 which use life forms as indicators of environmental quality.

U.S. Council on Environmental Quality. ENVIRONMENTAL QUALITY. Washington, D.C.: U.S. Government Printing Office, 1970-- . Annual.

> Annual report of the council gives an overview of the environment for the year including costs from pollution and abatement, laws, regulations, present status, trends, and perspectives. Contains general statistics.

U.S. Geological Survey. Office of Water Data Coordination. CATALOG OF INFORMATION ON WATER DATA. 4 vols. Washington, D.C.: U.S. Government Printing Office, 1968.

> Listing of the types of data collected by federal, state, and local agencies, and how to obtain that data. Areas covered include water quality, surface water, ground water, and areal investigations.

WATER QUALITY CRITERIA DATA BOOK. 4 vols. Water Pollution Control Research Series. Washington, D.C.: U.S. Environmental Protection Agency, 1970-72. Distributed by U.S. Government Printing Office.

> Series of volumes, each with its own author(s), covering organic pollution of freshwater, inorganic chemical pollution of freshwater, effects of chemicals on aquatic life, and recreational water quality. Most of each volume is devoted to tables which list the pollutant, its source and concentrations in the environment, as well as toxicity data, comments, and references.

Chapter 7
DIRECTORIES

Although most of the directories listed below were published in the last five years, the information they contain can become quickly dated. Current information can be obtained from several of the directories which are published yearly. Two journals, ENVIRONMENTAL SCIENCE AND TECHNOLOGY and WATER AND SEWAGE WORKS, and the yearly index, ENVIRONMENT INDEX, all publish annual directories. These are particularly useful in keeping up with the rapid turnover in personnel in governmental agencies. Scientific societies also publish membership lists.

Agricultural Sciences Information Network. DIRECTORY OF INFORMATION RESOURCES IN AGRICULTURE AND BIOLOGY. Beltsville, Md.: National Agricultural Library, 1971. Distributed by U.S. Government Printing Office. 523 p.

> Information is provided concerning the staff, scope, research areas and publications, plus funding and degrees offered at universities and units of land grant colleges.

American Institute of Biological Services. DIRECTORY OF BIOSCIENCE DEPARTMENTS IN THE UNITED STATES AND CANADA. New York: Reinhold Publishing, 1967. 672 p.

> Provides names and addresses of accredited academic institutions, the composition of their departments, courses offered, and requirements. Indicates if they have field stations.

AMERICAN MEN AND WOMEN OF SCIENCE: THE PHYSICAL AND BIOLOGICAL SCIENCES. 13th ed. 7 vols. New York: Jaques Cattel Press/ R.R. Bowker, 1977.

> Supplies biographical information plus the field of interest and address of people in science. Comparable to a "who's who" in science. Formerly AMERICAN MEN OF SCIENCE.

Directories

Cate, William, ed. DIRECTORY OF ENVIRONMENTAL CONSULTANTS. St. Louis: Directory Press, 1972. 46 p.

Names and addresses of consultants, listed by specialty.

Clark, William E., ed. CONSERVATION DIRECTORY. Washington, D.C.: National Wildlife Federation, 1974. x, 206 p.

Names, addresses, officers and staff, publications, and scope of organizations, agencies, or individuals concerned with conservation.

CONSULTANTS AND CONSULTING ORGANIZATIONS DIRECTORY. 4th ed. Edited by Paul Wasserman and Janice McLean. Detroit: Gale Research Co., 1979. xiv, 1,136 p.

Names and addresses of people and organizations engaged in consulting, as well as staff and scope of activity.

DIRECTORY OF INFORMATION RESOURCES IN THE UNITED STATES: BIOLOGICAL SCIENCES. Washington, D.C.: Library of Congress, National Referral Center, 1972. Distributed by U.S. Government Printing Office. iv, 577 p.

Titles, addresses, area of interest, publications, and so on of professional societies, university research bureaus and institutes, libraries, federal, and state agencies.

DIRECTORY OF INFORMATION RESOURCES IN THE UNITED STATES: WATER. Washington, D.C.: Library of Congress, National Referral Center, 1966. Distributed by U.S. Government Printing Office. v, 248 p.

Describes numerous organizations and institutions which are conducting research or collecting data on water and water-related subjects. Focuses primarily on freshwater.

DIRECTORY OF WATER POLLUTION RESEARCH LABORATORIES. Publication no. 18313. Paris: Organization for Economic Cooperation and Development, 1966. 519 p.

Directory information on research laboratories includes organization, personnel (type and number), documentation, special facilities and equipment, and research program. Worldwide in scope.

Gough, Beverly E., ed. WORLD ENVIRONMENTAL DIRECTORY. Silver Spring, Md.: Business Publishers, 1974-- . Irregular.

Address and area of interest are provided for a variety of groups with environmental concerns. Listings for the United States include: manufacturers of water pollution control products, professional services involving consulting, design and research, federal, state, and local government agencies, independent agencies, law firms, professional, scientific, technical, trade, labor and public interest organizations. Directory lists State Water Resources Research Institutes, analytical laboratories, universities and other educational institutions, environmental libraries and periodical publications. Also included are the officials in major industries who are in charge of environmental quality.

Hunsaker, D. II, and Dalgleish, R. BIOLOGICAL FIELD STATIONS OF NORTH AMERICA. New York: MSS Educational Publishing, 1970. 142 p.

Information obtained from a questionnaire in 1968-69. Includes address, staff names, facilities, and research in progress.

Institute of Ecology. DIRECTORY OF ENVIRONMENTAL LIFE SCIENTISTS. 9 vols. Washington, D.C.: U.S. Army Corps of Engineers, 1974. Distributed by U.S. Government Printing Office.

Each volume covers a different geographical region of the United States, and contains the names and addresses of environmental life scientists in that region, plus information on their principal research field, competence, experience, research activity, and geographical familiarity.

Kerbec, M.J., ed. YOUR GOVERNMENT AND THE ENVIRONMENT. 2 vols. Arlington, Va.: Output Systems, 1971-72.

Presents an overview of each area of environmental pollution, as well as a listing of organizations in federal and state government concerned with the environment. Includes standards, budgets, and, a glossary.

Onyx Group, Inc. ENVIRONMENT U.S.A.: A GUIDE TO AGENCIES, PEOPLE AND RESOURCES. New York: R.R. Bowker, 1974. xii, 451 p.

Listing of groups involved in environmental problems at federal, state, and private level. Includes a short description of purpose, address, and people to contact. Also included are chapters on fund raising, the law, lists of conferences, and films, and a glossary

Quigg, P[hilip].W., ed. WORLD DIRECTORY OF ENVIRONMENTAL EDUCATIONAL PROGRAMS: POST-SECONDARY STUDY AND TRAINING IN 70 COUNTRIES. New York: R.R. Bowker, 1973. xiii, 289 p.

Address of institution, cost of program, scholarships, and description of areas covered are given for more than one thousand programs.

Directories

SCIENTIFIC, TECHNICAL AND RELATED SOCIETIES OF THE UNITED STATES. 9th ed. Washington, D.C.: National Academy of Sciences, 1971. 213 p.

Provides address, history, purpose, membership, meetings, activities, and publications of societies in the United States.

Syracuse University Research Corporation Policy Institute. ENVIRONMENTAL RESEARCH LABORATORIES IN THE FEDERAL GOVERNMENT: AN INVENTORY. 2 vols. Syracuse, N.Y.: 1971.

Indicates what facilities are available, the type of research being conducted, the budget, and the staff.

Thibeau, C.E., ed. N.F.E.C. DIRECTORY OF ENVIRONMENTAL INFORMATION SOURCES. 2d ed. Boston: National Foundation for Environmental Control, 1972. Distributed by Canners Books. ix, 457 p.

Contains a wide range of information including a listing of U.S. governmental agencies, citizen's organizations, professional, trade, and educational societies, concerned with the environment, plus listings of abstracts, bibliographies, conferences, books, and films.

Troyer, D.L.; Kellogg, M.G.; and Anderson, H.O. SOURCE BOOK FOR BIOLOGICAL SCIENCES. New York: Macmillan, 1972. ix, 176 p.

A directory for teachers of the publishers, teaching materials, and books in biology including environmental science.

Trzyna, T., ed. DIRECTORY OF CONSUMER PROTECTION AND ENVIRONMENTAL AGENCIES. Orange, N.J.: Marquis Academic Media, 1973. xiii, 627 p.

Names, addresses, key personnel, purpose and activities, and publications of organizations involved in influencing, formulating, or administering consumer protection and environmental policy.

Trzyna, T.; Coan, E.V.; Rambach, P.; and Weiss, C., eds. WORLD DIRECTORY OF ENVIRONMENTAL ORGANIZATIONS. Claremont, Calif.: Sierra Club, 1973. v, 155 p.

Describes over sixteen hundred environmentally oriented organizations in two hundred countries, provides the name, address, and scope of the organization.

U.S. Council on Environmental Quality. FEDERAL ENVIRONMENTAL MONITORING DIRECTORY. Washington, D.C.: 1973. Distributed by U.S. Government Printing Office. xii, 105 p. Paperback.

Lists sources of primarily federal environmental monitoring information. Includes agency title, scope of activity, and person to contact.

U.S. Environmental Protection Agency. GROUPS THAT CAN HELP: A DIRECTORY OF ENVIRONMENTAL ORGANIZATIONS. Washington, D.C.: 1972. Distributed by U.S. Government Printing Office, 12 p. Pamphlet.

> Lists organizations, including their address and scope, which can be contacted by people seeking information about conservation.

United States International Environmental Referral Center. U.S. DIRECTORY OF ENVIRONMENTAL SOURCES. Washington, D.C.: U.S. Environmental Protection Agency, 1977. Distributed by U.S. Government Printing Office. 540 p.

> Lists names, addresses and type of information available from 1,114 sources in the United States, including 225 sources of information on water pollution. Federal agencies, state and local governments, universities, business and industry, societies and associations, laboratories, information center, consultants, and engineering firms are among the organizations registered with the USIERC and included in this directory.

WHO IS PUBLISHING IN SCIENCE. Philadelphia: Institute for Scientific Information, 1967-- . Annual.

> Cumulation of names and addresses of the first authors of papers published in journals indexed by CURRENT CONTENTS, SCIENCE CITATION INDEX, and SOCIAL SCIENCES CITATION INDEX.

WHO'S WHO IN CONSULTING. 2d. ed. Edited by Paul Wasserman et al. Detroit: Gale Research Co., 1973. xvii, 1,011 p.

> Names, addresses, biographical information, and publications of consultants for business, industry, and government.

Wilson, William, K.; Dowd, Morgan D.; and Sholtys, Phyllis A., eds. WORLD DIRECTORY OF ENVIRONMENTAL RESEARCH CENTERS. 2d ed. New York: Onyx Press, 1974. Distributed by R.R. Bowker. xi, 330 p.

> Names and addresses of more than forty-eight hundred research centers, worldwide.

Wolff, Garwood R., ed. ENVIRONMENTAL INFORMATION SOURCES HANDBOOK. New York: Simon and Schuster, 1974. 568 p.

> Directory covers sources of environmental information from both governmental agencies (federal, state and interstate), civic and conservation groups, professional societies, industry associations, information services, universities and regional organizations. Information provided includes address, person to contact, a description of purpose, and any regular publications and/or services available. Also listed are commercial newsletters, trade and business magazines.

Directories

WORLD GUIDE TO SCIENTIFIC ASSOCIATIONS. New York: R.R. Bowker, 1974. xii, 481 p.

Directory information on more than ten thousand scientific associations, organizations, and academies in 134 countries.

Young, Margaret Labash, and Young, Harold Chester. DIRECTORY OF SPECIAL LIBRARIES AND INFORMATION CENTERS. 5th ed. 3 vols. Detroit: Gale Research, 1979.

Names of libraries and sponsoring organizations as well as information on holdings, services, and publications.

Chapter 8
LEXICONS

Abercrombie, M.; Hickman, C.J.; and Johnson, M.L. A DICTIONARY OF BIOLOGY. 6th ed. Baltimore: Penguin Books, 1973. 311 p. Paperback.

Collocott, T.C., ed. DICTIONARY OF SCIENCE AND TECHNOLOGY. Edinburgh: W. and R. Chambers, 1971. xvi, 1,328 p.

Cover title reads CHAMBERS DICTIONARY OF SCIENCE AND TECHNOLOGY. Covers many fields including those encompassed by water pollution.

Crowley, Ellen T., ed. ACRONYMS, INITIALISMS, AND ABBREVIATIONS DICTIONARY: A GUIDE TO ALPHABETIC DESIGNATIONS, CONTRACTIONS, ACRONYMS, INITIALISMS, ABBREVIATIONS, AND SIMILAR CONDENSED APPELLATIONS. 6th ed. Detroit: Gale Research Co., 1978 xv, 635 p.

Primarily translation of acronyms.

Durrenberger, R.W., ed. DICTIONARY OF THE ENVIRONMENTAL SCIENCES. Palo Alto, Calif.: National Press Books, 1973. iv, 282 p.

Short, concise definitions of words from many disciplines.

Gecker, J.R.; Mackenthun, K.M.; and Ingram, W.M. GLOSSARY OF COMMONLY USED BIOLOGICAL AND RELATED TERMS IN WATER AND WASTE-WATER CONTROL. Publication no. 999-WP-2. Cincinnati: Public Health Service, 1963. iii, 22 p.

Definitions of aquatic and pollution biology terms. Includes numerous helpful illustrations.

Gilpin, Alan. DICTIONARY OF ENVIRONMENTAL TERMS. St. Lucia, Queensland: University of Queensland Press, 1976.

Terms included largely concerned with pollution and its control.

Lexicons

GLOSSARY: WATER AND WASTEWATER CONTROL ENGINEERING. New York: American Public Health Association, 1969. xi, 387 p.

For people in water and wastewater engineering.

Hackh, I.W.D. HACKH'S CHEMICAL DICTIONARY. 4th ed., completely rev. and edited by Julius Grant. New York: McGraw-Hill, 1969. xi, 738 p.

Definitions of words generally used in chemistry and related sciences.

Hanson, H.C. DICTIONARY OF ECOLOGY. New York: Philosophical Library, 1962. 382 p.

Includes references used in compiling the dictionary.

Henderson, I.F., and Henderson, W.D. A DICTIONARY OF BIOLOGICAL TERMS. 8th ed., edited by J.H. Kenneth. New York: Van Nostrand, 1963. xvi, 640 p.

Howell, J.V. GLOSSARY OF GEOLOGY AND RELATED SCIENCES. 2d ed. Washington, D.C.: American Geological Institute, 1960. x, 325 p.

Geological definitions, in many cases, cited from literature sources.

Jaeger, E.C. THE BIOLOGIST'S HANDBOOK OF PRONUNCIATIONS. Springfield, Ill.: Charles C Thomas, 1960. xvi, 317 p.

Gives pronunciation and in some cases, short (one to three words) definitions.

_____. SOURCEBOOK OF BIOLOGICAL NAMES AND TERMS. 3d ed. Springfield, Ill.: Charles C Thomas, 1955. xxxv, 317 p.

Definitions of word roots and combining forms plus examples of how the words are used.

Nelson, A., and Nelson, K.D. DICTIONARY OF WATER AND WATER ENGINEERING. Cleveland: CRC Press, 1973. vi, 270 p.

U.S. Bureau of Reclamation. THESAURUS OF WATER RESOURCES TERMS: A COLLECTION OF WATER RESOURCES AND RELATED TERMS FOR USE IN INDEXING TECHNICAL INFORMATION. Washington, D.C.: 1971. Distributed by U.S. Government Printing Office. vi, 339 p.

Used for locating words with similar meanings for literature searches or writing.

U.S. Environmental Protection Agency. GLOSSARY OF WATER RESOURCE TERMS. Chicago: 1970. 39 p.

Short, basic definition of many terms related to water pollution.

Veatch, J.O., and Humphrys, C.R. WATER AND WATER USE TERMINOLOGY. Kaukauna, Wis.: Thomas Printing and Publishing, 1966. 375 p.

Glossary of terms related to lakes, standing waters, and wetland. Includes geologic, economic, hydrologic, limnologic, ecologic, and engineering terms.

Chapter 9

ENCYCLOPEDIAS AND ATLASES

Fairbridge, R.W., ed. THE ENCYCLOPEDIA OF GEOCHEMISTRY AND ENVIRONMENTAL SCIENCES. Vol. 4, Section A. New York: Van Nostrand Reinhold, 1972. xxi, 1,321 p.

> Good source of information or aquatic chemistry, includes the various cycles and discussion of chemicals. Contains an overview of environmental pollution.

Geraghty, J.J.; Miller, D.W.; Leeden, F. van der; and Troise, F.L. WATER ATLAS OF THE UNITED STATES. 3d ed. Port Washington, N.Y.: Water Information Center, 1973. Unpaged.

> Contains 122 maps of the United States showing water quality, resources, use, and pollution. Included in the last section are thermal pollution, mercury, sewage, offshore disposal, and fish kills.

Gresswell, R.K., and Huxley, A. STANDARD ENCYCLOPEDIA OF THE WORLD'S RIVERS AND LAKES. New York: G.P. Putnam's Sons, 1965. 384 p.

> Contains basic information as to source, length, and tributaries of many rivers and lakes. Somewhat dated.

Lapedes, D.N., ed. McGRAW-HILL ENCYCLOPEDIA OF ENVIRONMENTAL SCIENCE. New York: McGraw-Hill, 1974. 754 p.

> Encyclopedia for the nonspecialist giving brief descriptions of various aspects of the environment including water pollution.

McGRAW-HILL ENCYCLOPEDIA OF SCIENCE AND TECHNOLOGY. 15 vols. New York: McGraw-Hill, 1971.

> General encyclopedia of scientific terms and principals. Section on water pollution includes analysis, purification, and pollution. Includes a yearbook which reports the latest findings in science.

Encyclopedias and Atlases

Todd, D.K., ed. THE WATER ENCYCLOPEDIA. Port Washington, N.Y.: Water Information, 1970. x, 559 p.

Sourcebook for facts related to water (physical and chemical), water use, and water pollution. Includes agencies (state, federal), as well as universities associated with water research and management.

Chapter 10

JOURNALS AND NEWSLETTERS

There are few journals restricted only to articles on water pollution, while conversely, many journals will occasionally include articles pertaining to water pollution. The best way to decide which journals suit your purpose is to consult the abstracts, indexes, and bibliographies listed in previous sections.

The following list of journals are those which generally contain articles on water pollution or areas closely related to the topic. The list has been divided into six sections indicating the scope of the journals.

A. WATER POLLUTION

JOURNAL WATER POLLUTION CONTROL FEDERATION. Washington, D.C.: Water Pollution Control Federation, 1928-- . Monthly.

> Articles range from water treatment to the detection, analysis and effects of water pollutants. June issue contains an annual review of the literature on various areas of water pollution. Each issue contains a list of meetings and a W.P.D.C. publications list. Formerly SEWAGE WORKS JOURNAL, 1928-50, SEWAGE AND INDUSTRIAL WORKS, 1950-60.

WATER, AIR AND SOIL POLLUTION. Boston: D. Reidel Publishing, 1972-- . Vol. 4, 1975. Quarterly.

> Articles on wide range of topics including the dispersion, occurrence, determination, removal, and effects of pollutants in air, water, and soil.

WATER RESEARCH. Elmsford, N.Y.: Pergamon Press, 1967-- . Monthly.

> Articles on water quality management, treatment, removal, detection, and effects of pollutants. Partially supersedes AIR AND WATER POLLUTION (1958-66).

WATER RESOURCES BULLETIN. Minneapolis: American Water Resources Association, 1965-- . Bimonthly.

Articles on hydrology, planning, and water. Analysis.

B. ENVIRONMENT

AMBIO. Stockholm: Royal Swedish Academy of Sciences, 1972-- . Bimonthly.

Includes articles on all aspects of environment, examples primarily European. Journal is in English.

CRITICAL REVIEWS IN ENVIRONMENTAL CONTROL. Cleveland: CRC Press, 1970-- . Quarterly.

Each issue contains several long review articles on some aspect of the control of an environmental contaminant or general problem area.

ENVIRONMENT. St. Louis: Scientist's Institute for Public Information, 1970-- . Monthly except January-February and July-August.

General, popular science type articles on various aspects of environmental pollution. Formerly SCIENTIST AND CITIZEN, 1958-69.

ENVIRONMENTAL AFFAIRS. Brighton, Mass.: Environmental Law Center, 1971-- . Quarterly.

Articles on legal and economic aspects of pollution.

ENVIRONMENTAL POLLUTION. Essex, Engl.: Applied Science Publishers, 1970-- . 8 times per year.

Articles on the biological effects of pollution.

ENVIRONMENTAL SCIENCE AND TECHNOLOGY. Easton, Pa.: American Chemical Society, 1967-- . Monthly.

Articles on many aspects of the environment. November issue is a pollution control directory, listing dealer and supply houses and professional consulting services, plus a list of books reviewed in the journal during the previous year.

JOURNAL OF ENVIRONMENTAL QUALITY. Madison, Wis.: American Society of Agronomy, Crop Science Society of America and Soil Science Society of America, 1972-- . Quarterly.

Articles cover all aspects of environmental quality in natural and agricultural ecosystems.

THE SCIENCE OF THE TOTAL ENVIRONMENT. Amsterdam: Elsevier Scientific Publishing, 1972-- . Quarterly.

Articles cover all aspects of man's effect on the environment, from analysis of pollutants to measurement of toxicity and effects on plant and animal communities.

C. WATER MANAGEMENT AND TREATMENT

AMERICAN WATER WORKS ASSOCIATION JOURNAL. Denver: American Water Works Association, 1914-- . Monthly.

Articles on management of water treatment facilities, regulations, and water quality and treatment. Each issue contains selected bibliographies, plus a list of AWWA publications, and lists of films and libraries.

JOURNAL OF THE ENVIRONMENTAL ENGINEERING DIVISION. Proceedings of the American Society of Civil Engineers. New York: American Society of Civil Engineers, 1873-- . Bimonthly.

Articles on water treatment and pollution. Formerly PROCEEDING AMERICAN SOCIETY OF CIVIL ENGINEERS, 1873-1956 and JOURNAL SANITARY ENGINEERING DIVISION, 1956-73.

WATER AND SEWAGE WORKS. Chicago: Water and Sewage Works, 1890-- . Monthly.

Articles concerned with water supply, and treatment, as well as wastewater treatment and management. Extra issue in April contains directory of water pollution control agencies and organizations, equipment, list of films, books, symposia, and courses.

D. PESTICIDES AND TOXICOLOGY

ARCHIVES OF ENVIRONMENTAL CONTAMINATION AND TOXICOLOGY. New York: Springer-Verlag, 1973-- . Quarterly.

Articles cover all aspects of contaminants in the environment with emphasis on the effect of contaminants in organisms and plants.

BULLETIN OF ENVIRONMENTAL CONTAMINATION AND TOXICOLOGY. New York: Springer-Verlag, 1966-- . Monthly.

Articles on air, soil, water, and food contamination and pollution as well as methodology.

ENVIRONMENTAL RESEARCH. New York: Academic Press, 1967-- . Bimonthly.

Publishes articles on biomedical ecology; primarily terrestrial.

PESTICIDES MONITORING JOURNAL. Washington, D.C.: Technical Services Division, U.S. Environmental Protection Agency, 1967-- . Quarterly.

Articles on pesticides and pesticide residues in relation to man and the environment.

RESIDUE REVIEWS. New York: Springer-Verlag, 1962-- . Irregular.

Reviews articles on all aspects of pesticides and other chemical residues; includes toxicological considerations.

E. HYDROLOGY AND AQUATIC ECOLOGY

ESTUARIES. Solomons, Md.: Estuarine Research Federation, Chesapeake Biological Laboratory, 1960-- . Quarterly.

Articles covering research and management studies on temperate estuaries and estuarine resources. Prior to 1978 called CHESAPEAKE SCIENCE.

ESTUARINE AND COASTAL MARINE SCIENCE. New York: Academic Press, 1973-- . Irregular.

Articles on all aspects of biological and physical phenomena associated with coastal and estuarine ecosystems.

FRESHWATER BIOLOGY. London: Blackwell Scientific Publications, 1970-- . Bimonthly.

Articles primarily on freshwater ecology and biology.

GROUND WATER. Columbus, Ohio: Water Well Journal Publishing, 1963-- . Bimonthly.

Articles on ground water hydrology.

HYDROBIOLOGIA. The Hague, Netherlands: Dr. W. Junk b.v. Publishers, 1948-- . Monthly.

Articles on pure and applied hydrobiology.

INTERNATIONAL ASSOCIATION OF THEORETICAL AND APPLIED LIMNOLOGY (Verhandungen Internationale Vereinigung für Theoretische und Angewandte Limnologie). Stuttgart: E. Schweizerbart'sche Verlagsbuchandlung, 1922-- . Triennial.

Proceedings of conferences held every three years including all aspects of aquatic ecology covering water analysis and pollution.

JOURNAL OF THE FISHERIES RESEARCH BOARD OF CANADA. Ottawa: Information Canada, 1919-- . Monthly.

Articles on fisheries management, limnology and oceanography. Formerly: CONTRIBUTION TO CANADIAN BIOLOGY, 1919-26; CONTRIBUTIONS TO CANADIAN BIOLOGY AND FISHERIES, 1926-34; JOURNAL OF THE BIOLOGICAL BOARD OF CANADA, 1934-37.

LIMNOLOGY AND OCEANOGRAPHY. Lawrence, Kans.: American Society of Limnology and Oceanography, 1956-- . Bimonthly.

Articles on all aspects of limnology and oceanography.

TRANSACTIONS OF THE AMERICAN FISHERIES SOCIETY. Lawrence, Kans.: Allen Press, 1870-- . Quarterly.

Papers on fish and fisheries, including ecology and the response of organisms to pollutants.

WATER RESOURCES RESEARCH. Washington, D.C.: American Geophysical Union, 1965-- . Bimonthly.

Articles on research in the social and natural sciences of water.

F. NEWSLETTERS

AIR/WATER POLLUTION REPORT. Silver Spring, Md.: Business Publishers, 1963-- . Weekly.

Short reports on environmental pollution, the law, markets, technology, available grants, and contracts as well as new publications. Absorbed AIR AND WATER NEWS in 1974.

CLEAN AIR AND WATER NEWS. Chicago: Commerce Clearing House, 1969-- . Weekly.

Contains news items, federal, and state regulations related to pollution, and new developments, plus lists of publications, symposia, and classes. Continued WATER CONTROL NEWS, 1966-68.

ENVIRONMENTAL REPORT. Washington, D.C.: Trends Publishing, 1970-- . Semimonthly.

Contains news items, new developments and a booklist.

Journals and Newsletters

MARINE POLLUTION BULLETIN. Elmsford, N.Y.: Maxwell House, 1970-- . Monthly.

> Publishes news, comments and reviews, as well as research articles on marine management, productivity, and pollution.

WATER NEWSLETTER. Port Washington, N.Y.: Water Information Center, 1958-- . Semimonthly. Supplement: RESEARCH AND DEVELOPMENT NEWS, 1959-- . Monthly.

> Newsletter contains short news items on government decisions, pollution research and what is being done by governmental agencies and industry. The supplement lists general interest water pollution reports, product development, and meetings. Dates, names, and addresses are included.

Chapter 11
FILMS

Pollution is a very popular topic which is reflected in the number of films on water pollution. This section includes both guides to films, as well as a list of selected films. Together with the guides listed below, films are included in ENVIRONMENT ABSTRACTS, ENVIRONMENT INDEX, and the AMERICAN WATER WORKS ASSOCIATION JOURNAL, plus the special annual issue of WATER AND SEWAGE WORKS.

A. GUIDES TO FILMS

AAAS SCIENCE FILM CATALOG. Ann Arbor, Mich.: R.R. Bowker, 1975. 398 p.

> Lists films in all areas of pure, applied, and social sciences for all educational levels, plus source and cost.

Canadian National Committee. WATER FILMS. 2d ed. Ottawa: Secretariat, Canadian National Committee for International Hydrological Decade, 1970. 194 p.

> Films on all aspects of water, including hydrology, water quality, and conservation. Films are available both in the United States and Canada.

ENVIRONMENT FILM REVIEW. New York: Environment Information Center, 1972-- . Annual.

> Critical review of films, plus name and address of distributors.

Specher, D. GUIDE TO FILMS (16mm) ABOUT ECOLOGY, ADAPTATION AND POLLUTION. Alexandria, Va.: Serina Press, 1971. iv, 55 p.

> Provides information as to film title, length, year, color, short description, source and, in some cases, educational level of audience.

B. FILMS

AH MAN, SEE WHAT YOU'VE DONE. 1971. North American Films, Box 919, Tarzana, Calif. 91356. 26 mins., color.

> Examines the impact of pollution on ocean plankton and dependent organisms. Presents a plea for man to stop polluting his world.

BEAUTIFUL RIVER. 1969. NBC Educational Enterprises (now distributed by Films Inc., 1144 Wilmette Avenue, Wilmette, Ill. 60091). 26 mins., color.

> Describes the fouling of the Connecticut River by industrial, municipal and thermal pollution. Winner: International Environmental Film Festival.

BRUSH CREEK BOUNCES BACK. 1970. Stuart Finley, 3428 Mansfield Road, Falls Church, Va. 22041. 22 mins., color.

> Illustrates the improved economic situation of a town in West Virginia after effective watershed management.

CHANGING RIVER, THE. N.d. International Film Bureau, 332 South Michigan Avenue, Chicago, Ill. 60604. 16 mins., color.

> Illustrates the relationship between aquatic organisms and their environment in an unpolluted river. It depicts a river from its beginning rivulets to its mouth.

THE CHOICE . . . IS YOURS. 1969. A-V Explorations, 505 Delaware Avenue, Buffalo, N.Y. 14202. 14 mins., color.

> Visual presentation of the many uses of water from recreation to a repository of pollutants. Problems caused by pollution are examined both in the laboratory and in water management agencies. American Film Festival Honors.

CONSERVING OUR WATER RESOURCES TODAY. N.d. Coronet Films, 65 East Water Street, Chicago, Ill. 60601. 11 mins., color, black and white.

> Examines the pressures that a growing population and expanding industrialization place on water resources. Indicates how water is used in the United States and the methods used in water conservation.

ECOLOGY OF FRESH WATER. 1967. Contemporary McGraw-Hill Films, 1221 Avenue of the Americas, New York, N.Y. 10020. 16 mins., color.

> Good introduction to the ecology of fresh water systems. The different kinds of aquatic organisms found in water are described and provide an understanding of the aquatic system in its unpolluted condition.

ESTUARINE HERITAGE. 1969. National Oceanic and Atmospheric Administration, U.S. Department of Commerce, 12231 Wilkins Avenue, Rockville, Md. 20852. 28 mins., color.

Depicts the effects of pollution of rivers on estuaries. The aquatic life in this area is also examined.

FROM THE RIDGE TO THE RIVER. N.d. N.A.C.D. Environmental Film Services, P.O. Box 855, League City, Tex. 77573. 26 mins., color.

Produced by the U.S. Department of Agriculture, film describes a watershed and its function. Effective watershed management is shown to decrease soil erosion and increase water storage.

THE GREAT LAKES: A MATTER OF SURVIVAL. N.d. BFA Educational Media, 2211 Michigan Avenue, Santa Monica, Calif. 90404. 11 mins., color.

Depicts the struggle to save the Great Lakes.

HOW TO MAKE A DIRTY RIVER. 1970. NBC Educational Enterprises (distributed now by: Films Incorporated, 1144 Wilmette Avenue, Wilmette, Ill. 60091). 27 mins., color.

Examines how the Passaic River in New Jersey became polluted and includes examples of resistance to pollution abatement and political problems encountered in the effort to clean up the river. Winner: Associated Press Best Local Documentary, International Environmental Film Festival, International Film and TV Festival Awards.

IT'S YOUR DECISION--CLEAN WATER. N.d. Association-Sterling Films, 866 Third Avenue, New York, N.Y. 10022. 14 mins., color.

Causes of water shortages are explored in this film along with an evaluation of the various kinds of waste treatment plants. Stresses the importance of treatment in wastewater reuse.

THE LAST MARSH. 1972. Films Incorporated, 1144 Wilmette Avenue, Wilmette, Ill. 60091. 10 mins., color.

Follows the last days of Powder Horn Marsh in Chicago, along with an illustration of the proper functions of a marsh.

OCEANS: THE SILENT CRISIS. N.d. Xerox Films, 245 Long Hill Road, Middletown, Conn. 06457. 24 mins., color.

Focuses on the sources of marine pollution, in this case, the Mississippi River. Various pollutants and pollution sources along the river are examined.

OIL ON THE RIVER. N.d. Stuart Finley, 3428 Mansfield Road, Falls Church, Va. 22041. 23 mins., color.

Presents the problems of waste management by the oil industry.

OUR POISONED WORLD: WATER. 1971. Time-Life Films, 43 West 16th Street, New York, N.Y.: 10011. 30 mins., color.

Discusses the effects of industrial pollution on Lake Michigan and other Michigan waterways.

THE PROBLEM WITH WATER IS PEOPLE. N.d. Contemporary McGraw-Hill Films, 1221 Avenue of the Americas, New York, N.Y. 10020. 30 mins., color.

Examines the problems associated with an increase in population, the accompanying increase in industrialization, and the need for irrigation waters with regard to the Colorado River.

THE RIVER MUST LIVE. 1966. Shell Film Library, 450 North Meridian Street, Indianapolis, Ind. 46204. 21 mins., color.

Describes the impact of water pollution on aquatic communities and how these organisms detoxify many pollutants. Stresses the need for better wastewater treatment.

THE RIVER THAT CAME BACK. N.d. Perennial Education, 1825 Willow Road, Northfield, Ill. 60093. 10 mins., color.

Depicts a successful conservation effort in the story of the AuSable River and stresses the need for continued vigilance.

ROOM TO BREATHE. 1969. NBC Educational Enterprises (now distributed by: Films Incorporated, 1144 Wilmette Avenue, Wilmette, Ill. 60091). 26 mins., color.

Explores the social and economic problems associated with redevelopment of the Delaware River Valley. Winner: International Environmental Film Festival.

TEAMWORK ON THE POTOMAC. 1964. Stuart Finley, 3428 Mansfield Road, Falls Church, Va. 22041. 28 mins., color.

Describes the cleanup program for the Potomac River and estuary.

THREATENED TREASURES. N.d. Audience Planners, 208 South LaSalle Street, Chicago, Ill. 60604. 5 mins., color.

Depicts the problems of pollution and possible solutions.

WATER. 1960. Pan American Health Organization Pan American Sanitary Bureau, 525 23d Street, N.W., Washington, D.C. 20037. 15 mins., color.

> Presentation of the worldwide crisis developing from the need for clean water for growing populations and associated industry and agriculture. Stresses the need for universal cooperation.

WATER POLLUTION: CAN WE KEEP OUR WATER CLEAN? 1971. Journal Films, 909 West Diversity Parkway, Chicago, Ill. 60614. 13 mins., color.

> Examines several sources and effects of pollution on a mountain stream. Pollutants include: trash, sewage, thermal pollution, oil (boats), pesticides, and fertilizers.

WHO KILLED LAKE ERIE? 1969. NBC Educational Enterprises (now distributed by Films Incorporated, 1144 Wilmette Avenue, Wilmette, Ill. 60091). 51 mins., color.

> Describes pollutants and their effect on aquatic life in Lake Erie. Examines problems associated with cleanup, especially the cost. Winner: Cine Golden Eagle, Peabody, International Film Festival, Headliner Achievement Award.

Part II

WATER POLLUTION REFERENCE SOURCES

Listed in this section are references relating to particular pollutants, pollutant sources, water quality control, and the impact of water pollution on our social, economic, and political structure. Each chapter (except 12 and 14) is, in general, divided into three parts: the first contains abstracts and bibliographies relating to the specific area, the second contains references to books and overview type journal articles, and the third is a listing of journal articles. The journal articles were selected to provide both an introduction to more current research, and to indicate the types of journals containing articles on the specific subject.

Chapter 12
AQUATIC ECOLOGY AND HYDROLOGY

Hydrology describes the physical and chemical properties of water, while the principles of aquatic ecology characterize the relations between the aquatic biota (both plant and animal) and the physical, chemical, and biological components of the environment. An understanding of the causes, effects, and treatment of water pollution is based on a knowledge of both these disciplines. Thus, the fate of a pollutant, such as DDT, in a stream is affected by chemical interactions with components of the water and sediment. Further, these interactions are mediated by physical phenomena, such as temperature, light, and flow. Ultimately the biological components of the ecosystem accumulate the compound and/or its transformed products.

Amos, William H. THE INFINITE RIVER: A BIOLOGIST'S VISION OF THE WORLD OF WATER. New York: Random House, 1970. xiv, 269 p.

> Narrative of the life of and in a river. Describes the hydrologic cycle, from seawater evaporation to cloud formation and the progress of water from a rivulet to the sea again. Also describes the life that inhabits the river during its various stages. The last chapter discusses man's impacts.

Barnes, R.S.K. ESTUARINE BIOLOGY. The Institute of Biology's Studies in Biology, no. 49. London: Edward Arnold, 1974. iv, 76 p.

> Short, somewhat technical introduction to the physical and biological features of an estuary. The primary focus of the text is on the estuarine fauna.

Barnes, R.S.K., and Green J., eds. THE ESTUARINE ENVIRONMENT. London: Applied Science Publishers, 1972. xii, 133 p.

> Collection of papers presented at a symposium covering estuarine animals, salt marsh plants, sedimentation, chemical processes, and the monitoring and cleanup of the Thames estuary. Provides a good introduction, as well as a reference source for estuarine ecology.

Aquatic Ecology and Hydrology

Brown, Alison L. ECOLOGY OF FRESH WATER. Cambridge, Mass.: Harvard University Press, 1971. xi, 129 p.

> Introduction to fresh water ecology; the physical and biological environment is described as well as energy flows and community distributions.

Chorley, Richard J., ed. WATER, EARTH, AND MAN: A SYNTHESIS OF HYDROLOGY, GEOMORPHOLOGY AND SOCIOECONOMIC GEOGRAPHY. London: Methuen and Co., 1969. xix, 588 p.

> Compilation of short descriptive articles on various aspects of the hydrological cycle. Topics covered include erosion, sedimentation, drainage basin precipitation, evapotranspiration, runoff, and water use. Semitechnical introduction to the hydrological cycle.

Coker, Robert E. STREAMS, LAKES, PONDS. Chapel Hill: University of North Carolina Press, 1954. 327 p.

> Presents a general description of the various water forms, their chemical nutrients and basic life forms.

Ford, Richard, and Hazen, William E., eds. READINGS IN AQUATIC ECOLOGY. Philadelphia: W.B. Saunders, 1972. vi, 397 p.

> Collection of abridged articles relating to aquatic ecology, including some of the foundation works in ecology. Oriented toward students of ecology and includes articles on marine and freshwater ecology.

Franks, Felix, ed. WATER: A COMPREHENSIVE TREATISE. Vol. 1: THE PHYSICS AND PHYSICAL CHEMISTRY OF WATER. New York: Plenum Press, 1972. xx, 596 p.

> Technical treatise on the basic properties of the intermolecular nature of water.

Friedrich, Hermann. MARINE BIOLOGY: AN INTRODUCTION TO ITS PROBLEMS AND RESULTS. Translated by Gwynne Vevers. Seattle: University of Washington Press, 1969. xii, 474 p.

> Comprehensive survey of marine biology aimed at an advanced college level. Discusses the dependence of living organisms upon abiotic conditions.

Golterman, H.L. PHYSIOLOGICAL LIMNOLOGY: AN APPROACH TO THE PHYSIOLOGY OF LAKE ECOSYSTEMS. Developments in Water Science, 2. New York: Elsevier Scientific Publishing, 1975. xiv, 490 p.

Describes the interactions between the abiotic environment and the organisms inhabiting the open water of lakes, primarily algae. Discusses the various nutrient cycles and budgets as well as the transfer of energy through the food chain. Includes sections on the origin of lakes and the problems of water management.

Hutchinson, G. Evelyn. A TREATISE ON LIMNOLOGY. 2 vols. New York: John Wiley and Sons, 1957.

One of the most complete coverages of the field by a foremost authority. Volume one covers geography, physics, and chemistry. Volume two is an introduction to lake biology and the limnoplankton.

Hynes, H.B.N. THE ECOLOGY OF RUNNING WATERS. Toronto: University of Toronto Press, 1970. xxiv, 555 p.

Most comprehensive book on stream limnology available. Considerable emphasis is given to stream insects.

Kinne, Otto, ed. MARINE ECOLOGY: A COMPREHENSIVE INTEGRATED TREATISE ON LIFE IN OCEANS AND COASTAL WATERS. Vol. 1: ENVIRONMENTAL FACTORS. New York: Wiley-Interscience, 1970. 681 p.

Summary and evaluation of the response of salt and brackish-water organisms to variations in abiotic and biotic factors. Each section reviews a single topic and includes the response of the various types of organisms from bacteria to plants and animals to the particular factor.

Laas, William, and Beicos, S.S. THE WATER IN YOUR LIFE. New York: Popular Library, 1967. 190 p.

Popular account of water, its properties, the hydrologic cycle, and water quality.

Lvovitch, M.I. "World Water Balance (General Report)." In WORLD WATER BALANCE, Proceedings of the Reading Symposium, International Association of Scientific Hydrology, vol. 2, pp. 404-14. Gentbrugge, Belgium: UNESCO/IASH/WMO, 1972.

Presents the concept of world water balance and its importance is examined for estimating world water resources. Historical review of various estimates made by scientists for different parts of the hydrological cycle is given with relevant statistics.

Macan, Thomas T. FRESHWATER ECOLOGY. New York: John Wiley and Sons, 1963. x, 338 p.

Text is concerned with the limiting factors, i.e. physical, chemical, and biological found in a freshwater environment, and the interrelationships which exist.

McLusky, Donald S. ECOLOGY OF ESTUARIES. London: Heinemann Educational Books, 1971. viii, 144 p.

General, not overly technical, review of estuarine ecology. Discusses the physical parameters, the organisms which occur, and some of the adaptations for their survival.

Manahan, Stanley E. ENVIRONMENTAL CHEMISTRY. Boston: Willard Grant Press, 1972. v, 393 p.

Approximately half the text pertains to water chemistry and covers such topics as the role of microorganisms in chemical reactions, water analysis, treatment, and pollution.

Moore, Hilary B. MARINE ECOLOGY. New York: John Wiley and Sons, 1958. 493 p.

Textbook covering the physical, chemical, and biological environmental factors and their interactions.

Pereira, H.C. "The Influence of Man on the Hydrological Cycle." In WORLD WATER BALANCE, Proceedings of the Reading Symposium, International Association of Scientific Hydrology, vol. 3, pp. 553-69. Gentbrugge, Belgium: UNESCO/IASH/WMO, 1972.

General discussion of the effects of land use management on the hydrologic cycle.

Pryde, Lucy T. CHEMISTRY OF THE WATER ENVIRONMENT. Reprint of Chapter 6 of Pryde's ENVIRONMENTAL CHEMISTRY, pp. 171-231. Menlo Park, Calif.: Cummings Publishing, 1973.

Overview of water chemistry contains a glossary of terms and discusses water, water treatment, and water pollution.

Reid, George K., and Wood, Richard D. ECOLOGY OF INLAND WATERS AND ESTUARIES. New York: Reinhold Publishing, 1976. 485 p.

Text presents an introduction to the processes which operate in lakes, streams, and estuaries. The book is divided into five sections: origins of lake basins and stream channels, the nature of water, the physical and chemical characteristics of aquatic systems, the plants and animals adapted to these systems, and the principles of population and community organization.

Remane, Adolph, and Schlieper, Carl. BIOLOGY OF BRACKISH WATERS. 2d rev. ed. New York: Wiley-Interscience, 1971. viii, 372 p.

> Divided into two sections: ecology and physiology. First section concerns the interactions between communities and the effects of physical factors on the communities. Second section covers the physiological response of brackish water organisms to physical and chemical factors encountered in their environment. Authors are European authorities.

Ruttner, F. FUNDAMENTALS OF LIMNOLOGY. 3d ed. Translated by D.G. Frey and F.E.J. Fry. Toronto: University of Toronto Press, 1965. xvi, 295 p.

> One of the classical texts in limnology, although somewhat dated. Emphasis is given to physical and chemical factors.

Tait, Ronald V. ELEMENTS OF MARINE ECOLOGY: AN INTRODUCTORY COURSE. New York: Plenum Press, 1968. vii, 272 p.

> Introductory text which covers the various phyla and environmental factors present in the marine environment.

Tebbutt, T.H.Y. WATER SCIENCE AND TECHNOLOGY. London: John Murray, 1973. xiv, 195 p.

> General text on water, its characteristics, uses, treatment, and pollution. Somewhat technical but presents an overview of water from an engineering viewpoint. Most examples are British.

U.S. Department of Interior. RIVER OF LIFE, WATER: THE ENVIRONMENTAL CHALLENGE. Conservation Yearbook Series, no. 6. Washington, D.C.: U.S. Government Printing Office, 1970. 96 p.

> Illustrated, easy reading overview of the "story" of water, covering the hydrological cycle, water quality, water distribution, and utilization.

Vernberg, Winona B., and Vernberg, F. John. ENVIRONMENTAL PHYSIOLOGY OF MARINE ANIMALS. New York: Springer-Verlag, 1972. ix, 346 p.

> Presentation of the physiological mechanisms which allow marine organisms to adapt to their particular habitat. For each general habitat area, adaptations related to chemical regulation, circulation, reproduction, and development are discussed.

Wetzel, Robert G. LIMNOLOGY. Philadelphia: W.B. Saunders, 1975. xii, 743 p.

> Written for advanced undergraduate to graduate student, this text provides a discussion of all aspects of the freshwater environment with emphasis on integration of the physical, chemical and biological components.

Whitton, B.A., ed. RIVER ECOLOGY. Studies in Ecology, vol. 2. Berkeley and Los Angeles: University of California Press, 1975. 725 p.

Each area of river ecology is reviewed by an authority in that area. Primarily covers the biological features of a river, with a chapter devoted to each category of organisms, i.e., algae, bacteria, and fungi, macro-invertebrates, and so forth. Includes sections on water pollution and surveys of river systems.

Wood, E.J. Ferguson. MARINE MICROBIAL ECOLOGY. New York: Reinhold Publishing, 1965. xii, 243 p.

Describes the microbial organisms of the marine environment and their interrelationship with the physical and chemical, as well as the biological, environment. Included is a section on the role of microorganisms in the productivity of the sea. Other sections cover pathological microorganisms and methods in marine microbiology.

Chapter 13
METHODS AND PROCEDURES

In recent years there has been rapid advancement in pollution detection instrumentation. Where the limits of detection were once in the parts per thousand (ppt) (g/1) or ppm (mg/1) range, they are now ppb (mg/1) and lower. Advances in methods and instrumentation have also been stimulated by the introduction of new pollutants. While pollutants can now, in most cases, be accurately measured, the assessment of the effects of pollution on aquatic systems in still imprecise. The bioassay remains the standard method for measuring a compound's toxicity, with death as the indicator. A concentration much less than that causing mortality is chosen as safe. This method provides no measure of sublethal effects, such as reduced growth or reproductive potential, which may ultimately affect competition and community structure. The rigorous assessment of pollution awaits better methods for measuring sublethal levels and the integration of the principles of aquatic ecology in these methods.

Among the best single source for initiating a survey of this topic is STANDARD METHODS FOR THE EXAMINATION OF WATER AND WASTEWATER INCLUDING BOTTOM SEDIMENTS AND SLUDGES (American Public Health Association, 1975). Each year the June issue of the JOURNAL WATER POLLUTION CONTROL FEDERATION (see p. 47) contains an excellent review of new methods and analyses relating to water pollution.

The journal articles at the end of this section are indicative of the journals containing literature on methodology. They also provide access to the more recent literature, especially if used in conjunction with SCIENCE CITATION INDEX (see p. 5).

A. ABSTRACTS AND BIBLIOGRAPHIES

Sinha, Evelyn. METHODS, MODELS AND INSTRUMENTATION FOR STUDIES OF AQUATIC POLLUTION: AN ANNOTATED BIBLIOGRAPHY. Ocean Engineering Information Series, vol. 5. La Jolla, Calif.: Ocean Engineering Information Service, 1971. iv, 29 p.

Annotated bibliography of over two hundred citations (1968-70) which emphasizes procedures and equipment used in field and laboratory investigations of aquatic pollution, and in abatement.

B. GENERAL REFERENCES

American Public Health Association. STANDARD METHODS FOR THE EXAMINATION OF WATER AND WASTEWATER INCLUDING BOTTOM SEDIMENTS AND SLUDGES. 14th ed. Edited by Arnold Greenberg et al. New York: 1975. 1,193 p.

Compilation of methods for the analysis of water from domestic and industrial origin, the best source of methods and the most widely used. Contains information on apparatus, reagents, procedures, calculations, and the best way to express the results. Attempts to standardize methods.

American Society for Testing and Materials. MICROORGANIC MATTER IN WATER: A SYMPOSIUM. ASTM Special Technical Publication, no. 448. Philadelphia: 1969. 124 p.

Devoted primarily to the analysis of organic contaminants in water such as organic pesticides, phenols, and molybdenum. Each paper describes the development of techniques for the analysis of a particular compound or class of compounds.

______. WATER QUALITY PARAMETERS. ASTM Special Technical Publication, no. 573. Philadelphia: 1975. 580 p.

Papers given at a symposium cosponsored by the Canadian Centre for Inland Waters and the Analytical Chemistry Division of the Chemical Institute of Canada. They cover a wide variety of analytical techniques for the determination of pollutants. Included are techniques for inorganic, organic, and biological analysis, monitoring, and remote sensing.

Benton, Allen H., and Werner, William E., Jr. MANUAL OF FIELD BIOLOGY AND ECOLOGY. 3d ed. New York: McGraw-Hill, 1974. 384 p.

Workbook of field methods indicating how to gather data on both the aquatic communities and on the physical and chemical parameters.

Burgess, Fred J., and James, Wesley P. AIR PHOTO ANALYSIS OF OCEAN OUTFALL DISPERSION. Water Pollution Control Research Series. Washington, D.C.: U.S. Environmental Protection Agency, Office of Research and Monitoring, 1971. Distributed by U.S. Government Printing Office. xi, 287 p.

Report on a study which utilized aerial photography to observe an ocean outfall for a period of several years. Waste concentrations, as well as diffusion and velocities were determined by a comparison with onsight sampling.

Cairns, John, Jr., and Dickson, K.L., eds. BIOLOGICAL METHODS FOR THE ASSESSMENT OF WATER QUALITY. ASTM Special Publication, no. 528. Philadelphia: American Society for Testing and Materials, 1973. vii, 256 p.

Papers presented at a symposium provide information on the types of considerations involved in making biological assessment. Indicate existing techniques, as well as some historic perspective. Topics include the use of fish, algae, aquatic invertebrates, and bacteria in the assessment of pollution, and bioassay techniques for toxicity evaluation.

Chapman, Robert L., ed. ENVIRONMENTAL POLLUTION INSTRUMENTATION. Pittsburgh: Instrument Society of America, 1969. viii, 167 p.

Fairly technical description of the various instruments used to measure pollution (both air and water). Contents include: the future of instrumentation in water pollution control, control system for paper mill effluent, and methods for measuring the concentration of radio isotopes in water.

Doudoroff, P[eter]., et al. "Bioassay Methods for the Evaluation of Acute Toxicity of Industrial Wastes to Fish." SEWAGE AND INDUSTRIAL WASTES 23 (November 1951): 1380-97.

Classical paper on bioassay technique for fish. Detailed instructions are given for the care of the fish and the setting up of concentrations, plus analysis of data.

Environmental Instrumentation Group, Lawrence Berkeley Laboratories, University of California. INSTRUMENTATION FOR ENVIRONMENTAL MONITORING. 4 vols. Berkeley: 1972-- . Looseleaf.

Continuing compilation of information on instrumentation used in environmental monitoring, includes type of instrument, calibration, sources and effects of contamination, characteristics, types available, and recommendations. Glossary is included. Volume 2 contains instrumentation suitable for the analysis of freshwater and brackish water (not marine).

Glass, Gary E., ed. BIOASSAY TECHNIQUES AND ENVIRONMENTAL CHEMISTRY. Ann Arbor, Mich.: Ann Arbor Science Publishers, 1973. xi, 499 p.

Compilation of various bioassay techniques for a wide variety of organisms and conditions. It contains the following sections: illustration of environmental complexities (a review article on mercury), uses of bioassay emphasizing general principles, examples of end points and indicators used in bioassay, conditions and supporting techniques for bioassay and applications.

Goldberg, Edward D., ed. A GUIDE TO MARINE POLLUTION. New York: Gordon & Breach Science Publishers, 1972. 168 p.

Expansion of the final report of the "Seminar on Methods of Detection, Measurement and Monitoring of Pollutants in the Marine Environment," sponsored by the Food and Agricultural Organization of the United Nations. Text is a compilation of synopses of methods for the detection, measurement, and monitoring of many marine pollutants. Included is a chapter on test organisms and a list of people and/or laboratories engaged in research.

Golterman, H.L., ed. METHODS FOR CHEMICAL ANALYSIS OF FRESH WATERS. IBP Handbook, no. 8. Oxford, Engl.: Blackwell Scientific, 1969. xvi, 172 p.

Designed as a guide for researchers interested in those biological processess in fresh water that lead, with interconversion of energy, to the production of organic material. Methods included are those which allow for a chemical characterization of fresh waters.

Golterman, H.L., and Clymo, R.S., eds. CHEMICAL ENVIRONMENT IN THE AQUATIC HABITAT. Amsterdam: N.V. Noord-Hollandshe Uitgevers Maatschappij, 1967. 322 p.

Proceedings of a IBP symposium (International Biological Programme) concerned with the routine analysis of such substances as organics, phosphorus, silica and iron, as well as the problems of nitrogen fixation, pH regulation, and the interaction of water and bottom sediments. Task of the conference was to produce a manual for the routine analysis of water (see reference above).

Kittrell, F.W. A PRACTICAL GUIDE TO WATER QUALITY STUDIES OF STREAMS. Washington, D.C.: Federal Water Quality Administration, 1969. Distributed by U.S. Government Printing Office. 135 p. Paperbound.

Meant as a guide to beginners, this manual described specific procedures peculiar to stream pollution studies. Topics include the physical characteristics of streams, sampling stations and procedures, sample examination, stream flows, time-of-water travel, and interpretation of data.

Klein, Louis. RIVER POLLUTION. Vol. 1: CHEMICAL ANALYSIS. London: Butterworth, 1959. ix, 206 p.

Covers the physical and chemical analysis of a stream. Includes dissolved oxygen, combined nitrogen, sulphur compounds, and metallic contaminants. Somewhat dated but a good source book.

Kolthoff, Izaak, M.; Elving, P.J.; and Stross, F.H. ANALYTICAL CHEMISTRY IN INDUSTRY. Vol. 2. Part III of TREATISE ON ANALYTICAL CHEMISTRY. New York: Wiley-Interscience, 1971. xxiv, 597 p.

Compilation of various analytical methods for waters from industry and treatment plants. Sections on quality standards and on chemical analyses commonly used for measuring acidity, alkalinity, chlorine, calcium, and aluminum.

Lee, G. Fred. "Chemical Aspects of Bioassay Techniques for Establishing Water Quality Criteria." WATER RESEARCH 7 (November 1973): 1525-46.

Discussion and review of the chemical processes which can effect compounds in water and thus alter the results of a laboratory-run bioassay.

Leithe, Wolfgang. ANALYSIS OF ORGANIC POLLUTANTS IN WATER AND WASTEWATER. Translated by STS, Inc. Ann Arbor, Mich.: Ann Arbor Science Publishers, 1973. ix, 213 p.

Detailed and technical presentation on the analysis of organics in water. Author first presents some general information on waste treatment and sampling techniques.

Mancy, Khalil H. INSTRUMENTAL ANALYSIS FOR WATER POLLUTION CONTROL. Ann Arbor, Mich.: Ann Arbor Science Publishers, 1971. xi, 331 p.

Emphasizes the application of instrumentation, including automatic and remote measurement systems to water quality characterization. Includes a discussion of the physical characteristics of water, as well as specific analytical techniques for a variety of compounds.

Martin, Dean F. MARINE CHEMISTRY. Vol. 1: ANALYTICAL METHODS. 2d ed. New York: Marcel Dekker, 1972. 389 p.

Prepared for the researcher without professional experience in the field. For each determination several experimental techniques are presented along with sections on reagents and equipment. Included are most of the commonly used methods for the analysis of compounds in marine waters, primarily nonpollutants.

Perkins, E.J. "Some Methods of Assessment of Toxic Effects Upon Marine Invertebrates." PROCEEDINGS OF SOCIETY FOR ANALYTICAL CHEMISTRY: ANALYTICAL DIVISION, CHEMISTRY 9 (May 1972): 105-14.

Discussion of the drawbacks in toxicity studies and proposals for bioassays procedures which would minimize these problems. While the study is based on marine invertebrates, it has wide application.

Rainwater, Frank H., and Thatcher, L.L. METHODS FOR COLLECTION AND ANALYSIS OF WATER SAMPLES. Geological Survey Water Supply Paper 1454. Washington, D.C.: U.S. Government Printing Office, 1960. ix, 301 p.

Indicates how to collect water samples, special materials needed to test for certain elements, and the instrumentation and analytical procedures necessary for the evaluation of water quality aspects from acidity to zinc.

Renn, Charles E. INVESTIGATING WATER PROBLEMS. Chestertown, Md.: LaMotte Chemical Products, 1970. 72 p.

Presentation of methods for the analysis of water quality. Includes a glossary and suggestions for the collection and preservation of samples.

Rodier, J. ANALYSIS OF WATER. Translated by IPST (Israel Program for Scientific Translations) Staff. New York: John Wiley and Sons, 1975. xvii, 926 p.

Comprehensive treatise on water analysis (includes fresh, salt water, and residual waters). Includes analysis for chemical and organic pollutants. Covers instrumentation, statistical analysis, standard solutions, and the biological quality of freshwaters. Technical.

Sawyer, Clair N., and McCarty, Perry L. CHEMISTRY FOR SANITARY ENGINEERS. 2d ed. New York: McGraw-Hill, 1967. xi, 518 p.

Textbook has two sections: the first on fundamentals of chemistry for engineers, and the second on water and wastewater analysis. Chapters on water analysis stress the basic chemistry of each analysis.

Sittig, Marshall. POLLUTANT REMOVAL HANDBOOK. Park Ridge, N.J.: Noyes Data, 1973. 528 p.

Compilation of methods on the removal of pollutants, particularly those emanating from industrial processes. Supplemental bibliography is provided.

Sprague, J.B. "Measurement of Pollutant Toxicity to Fish. I. Bioassay Methods for Acute Toxicity." WATER RESEARCH 3 (November 1969): 793-821. "II. Utilizing and Applying Bioassay Results." WATER RESEARCH 4 (January 1970): 3-32. "III. Sublethal Effects and 'Safe' Concentrations." WATER RESEARCH 5 (June 1971): 245-66.

Series of three articles provides a good introduction to the technique of bioassay by one of the authorities in the field. Includes terminology, as well as procedures for setting up a bioassay and interpreting the results. It also provides a critical review of the literature up to 1969.

Thomas, William A., ed. INDICATORS OF ENVIRONMENTAL QUALITY. Environmental Science Research, vol. 1. New York: Plenum Press, 1972. x, 275 p.

Proceedings of a symposium by the American Association for the Advancement of Science. Contents include discussions on the usefulness of environmental quality indexes, on establishing priorities, and on the various indexes used to monitor the environment.

Traversy, W.J. METHODS FOR CHEMICAL ANALYSIS OF WATERS AND WASTEWATERS. Ottawa: Canadian Department of Fisheries and Forestry, Water Quality Division, 1971. 169 p.

Water analysis manual in which analytical procedures are described for analyzing fifty-five water parameters.

U.S. Environmental Protection Agency. Water Quality Office. METHODS FOR CHEMICAL ANALYSIS OF WATER AND WASTES. Cincinnati: 1971. Distributed by U.S. Government Printing Office. xi, 312 p.

Compilation of analytical procedures for the chemical analysis of water and waste samples to be implemented by Water Quality Office laboratories.

C. JOURNAL ARTICLES

Arin, M.L. "Monitoring with Carbon Analyzers." ENVIRONMENTAL SCIENCE AND TECHNOLOGY 8 (October 1974): 898-902.

Brezonik, P.L., and Carriker, N.E. "Continuous Monitoring Automated Analysis and Sampling Procedures." JOURNAL WATER POLLUTION CONTROL FEDERATION 48 (June 1976): 1077-86.

Coleman, R.F. "Comparison of Analytical Techniques for Inorganic Pollutants." ANALYTICAL CHEMISTRY 46 (October 1974): 989A-96A.

Eberhardt, L.L. "Some Methodology for Appraising Contaminants in Aquatic Systems." JOURNAL FISHERIES RESEARCH BOARD OF CANADA 32 (October 1975): 1852-59.

Frant, M.S. "Detecting Pollutants with Chemical-Sensing Electrodes." ENVIRONMENTAL SCIENCE AND TECHNOLOGY 8 (March 1974): 224-28.

Jensen, K. "Hatching Rate as Bioassay Proposal for a Standard Technique." BULLETIN OF ENVIRONMENTAL CONTAMINATION AND TOXICOLOGY 14 (November 1975): 562-64.

Larimore, R.W. "Stream Drift as an Indicator of Water Quality." TRANSACTIONS OF AMERICAN FISHERY SOCIETY 103 (July 1974): 507-17.

Phillips, S.L., and Mack, D.A. "Instruments for Water Quality Measurement." ENVIRONMENTAL SCIENCE AND TECHNOLOGY 9 (March 1975): 214-20.

Walker, J.E., and Dahn, D.B. "Measuring Environmental Stress." ENVIRONMENTAL SCIENCE AND TECHNOLOGY 9 (August 1975): 714-19.

Especially informative on thermal pollution detection.

Wolman, M.G. "The Nation's Rivers." SCIENCE 174 (26 November 1971): 904-18.

Chapter 14
WATER POLLUTION—GENERAL

Relatively few textbooks have surfaced dealing with the broad topic of water pollution. Many authors have concentrated on selected aspects of the area of water pollution. Among the texts included in this section that provide the best overview of water pollution are Hynes (p. 77), Klein (p. 78), Warren (p. 82), Wilber (p. 82), and Mellanby (p. 80). Of these, Warren's book is somewhat technical and designed for the advanced reader, and Wilber's book is generally not well received by advanced students. Others providing information in a specific area or aspect are: Ciaccio (p. 76) handbook, Dugan (p. 77) biochemical, Hood (p. 77) marine, Jones (p. 78) fish, Vernberg and Vernberg (p. 82) physiology, Hart and Fuller (p. 77) freshwater invertebrate tolerance of pollutants, McCaull and Crossland (p. 79) legal and technological aspects, and Montgomery (p. 80) water pollution history. Additional references provide a popular view of water pollution or provide collections of papers on the subject.

Adler, C.A. ECOLOGICAL FANTASIES: DEATH FROM FALLING WATERMELONS. New York: Green Eagle Press, 1973. 350 p.

> Author looks at what he terms "persistent environmental myths" and attempts to separate ecological truth from fiction.

American Chemical Society. CLEANING OUR ENVIRONMENT: THE CHEMICAL BASIS FOR ACTION. Washington, D.C.: 1969. ix, 249 p.

> Nontechnical survey of the various areas of pollution concern (i.e., air, water, solid wastes, and pesticides) and an indication of what is known and what information is needed. Somewhat dated.

Ballentine, Robert K.; Reznek, S.R.; and Hall, C.W. SUBSURFACE POLLUTION PROBLEMS IN THE U.S. Technical Studies Report TS-00-72-2. Washington, D.C.: Environmental Protection Agency, Office of Water Programs, 1972. 24 p.

> General survey report on the sources of subsurface water contamination and the effect on water supplies.

Water Pollution—General

Bartleson, G.C. THE CHEMICAL INVESTIGATION OF RECENT LAKE SEDIMENTS FROM WISCONSIN LAKES AND THEIR INTERPRETATION. Water Pollution Control Research Series. Washington, D.C.: U.S. Environmental Protection Agency, 1971. 280 p.

> Results of a study which compared the postcultural and precultural sedimentation rates of elements.

Baxter, William F. PEOPLE OR PENGUINS: THE CASE FOR OPTIMAL POLLUTION. New York: Columbia University Press, 1974. 110 p.

> Pollution problem is set in economic and legal terms, as the author contends that "ecological aspects not important until the human cost is determined."

Brittin, Wesley E.; West, Ronald; and Williams, Robert, eds. AIR AND WATER POLLUTION. Boulder: Colorado Associated University Press, 1972. xvii, 613 p.

> Collection of papers given at a workshop interested in research needs in pollution control. Water pollution problems in two areas were covered: the physical and chemical phenomena in water treatment, and the effects of pollutants on the biota.

Bylinsky, Gene [T.]. "The Limited War on Water Pollution." In MAN'S IMPACT ON ENVIRONMENT, edited by Thomas R. Detwyler, pp. 195-204. New York: McGraw-Hill, 1971.

> General discussion of the causes of pollution and the problems encountered in the cleanup.

Catholic University of America. Institute of Ocean Science and Engineering. A LITERATURE SURVEY OF OCEAN POLLUTION. Washington, D.C.: 1971. 116 p.

> Surveys past studies on the problems of ocean pollution, chief pollutant sources, effects, and control. Also identifies areas of research needs.

Ciaccio, Leonard L., ed. WATER AND WATER POLLUTION HANDBOOK. 4 vols. New York: Marcel Dekker, 1971-73.

> Volume 1 contains papers on the physical, chemical and biological characteristics of waters and the effects of pollution on aquatic life. Volumes 2 through 4 are concerned with analysis and sampling of water, as well as chemical reactions in water, and instrumentation.

Cronin, L. Eugene. "The Role of Man in Estuarine Processes." In ESTUARIES, edited by G.H. Lauff, pp. 667-89. Washington, D.C.: American Association Advancement of Science, 1967.

Discussion of the various pollutants and polluting activities affecting the estuaries. Selected case studies are presented.

Detwyler, Thomas R., ed. MAN'S IMPACT ON ENVIRONMENT. New York: McGraw-Hill, 1971. xiii, 731 p.

Omnibus volume on pollution with eight articles specifically on water pollution including a general overview and discussions of specific pollutants, such as temperature and oil. Articles are reprints or excerpts from journals or books.

Dugan, Patrick R. BIOCHEMICAL ECOLOGY OF WATER POLLUTION. New York: Plenum Press, 1972. ix, 159 p.

General text which considers both the biochemistry of pollutants and the demand for water and water chemistry.

Faber, Harry A., and Bryson, Lena J., eds. CONFERENCE ON PHYSIOLOGICAL ASPECTS OF WATER QUALITY. Washington, D.C.: Public Health Service, 1960. v, 244 p.

Cross section of articles on the state-of-the-art, in the measurement of pollutants and their physiological effects on primarily nonaquatic organisms.

Hart, C.W., Jr., and Fuller, Samuel L.H., eds. POLLUTION ECOLOGY OF FRESHWATER INVERTEBRATES. New York: Academic Press, 1974. xiv, 389 p.

Collection of ten review-type articles on various freshwater invertebrates, from protozoa to mollusks and insects, indicating their ecology, usefulness as indications of pollution, and the effect of pollutants on them. Each paper covers the topic somewhat differently, but all provide a good overview of their particular organism with respect to pollution.

Hood, Donald W., ed. IMPINGMENT OF MAN ON THE OCEANS. New York: Wiley-Interscience, 1971. x, 738 p.

Book of technical papers which discuss the hydraulics and chemistry of the oceans, as well as the pollutants and their effect on the oceans. Sections on the coastal environment and modeling are also included.

Hynes, H.B.N. THE BIOLOGY OF POLLUTED WATERS. Liverpool, Engl.: Liverpool University Press, 1960. 202 p.

One of the best books on water pollution, although somewhat dated now. All the major topics related to pollution are included. Includes a well-selected bibliography for further reading.

Ingram, William M.; Mackenthun, Kenneth M.; and Bartsch, Alfred F. BIOLOGICAL FIELD INVESTIGATIVE DATA FOR WATER POLLUTION SURVEYS. Washington, D.C.: U.S. Department of Interior, Federal Water Pollution Control Administration, 1966. iv, 139 p.

> Introduction for the nonbiologist to the life sciences as they relate to water pollution. Also an introduction to field studies of polluted streams and lakes. Includes a section on terminology.

Jones, J.R. Erichsen. FISH AND RIVER POLLUTION. London: Butterworth, 1964. viii, 203 p.

> Provides information on the effect of major pollutants on fish. Pollutants covered include lead, zinc, copper, cyanides, acids, phenols, and so on. While somewhat dated, it does provide a good beginning reference.

Keup, Lowell E.; Ingram, William M.; and Mackenthun, Kenneth M., eds. BIOLOGY OF WATER POLLUTION: A COLLECTION OF SELECTED PAPERS ON STREAM POLLUTION, WASTE WATER, AND WATER TREATMENT. Cincinnati: U.S. Department of Interior, Federal Water Pollution Control Administration, 1967. 290 p.

> Compilation of papers, many of them classics, dealing with the effects of pollutants on aquatic organisms. Includes articles on the relationship of pollution to plankton, benthos, fish, and man. Includes a wide variety of articles.

Klein, Louis. RIVER POLLUTION. Vol. 2: CAUSES AND EFFECTS. London: Butterworth, 1962. v, 456 p.

> Text gives the causes of river pollution, the various types of pollution and their effect on the chemistry of the river and on the aquatic life (primarily fish).

Koeman, J.H., and Strik, J.J.T.W.A., eds. SUBLETHAL EFFECTS OF TOXIC CHEMICALS ON AQUATIC ANIMALS. New York: Elsevier Scientific Publishing, 1975. vi, 234 p.

> Proceedings of a symposium which covered descriptions of methods, test systems and test organisms, as well as laboratory and field studies concerning the sublethal effects of various metals, such as cadmium, zinc, mercury, and lead, and other chemicals, such as PCBs and dieldrin on fish, crustacea and other aquatic organisms.

Kruse, C.W. "Our Nation's Water: It's Pollution Control and Management." In ADVANCES IN ENVIRONMENTAL SCIENCE, vol. 1, edited by J.N. Pitts and R.L. Metcalf, pp. 41-71. New York: Wiley-Interscience, 1969.

> Overview of the history of water quality and wastewater treatment.

McCaull, Julian, and Crossland, Janice. WATER POLLUTION. Environmental Issues Series. New York: Harcourt Brace Jovanovich, 1974. xiv, 206 p.

Presents a nontechnical overview of water pollution and a critical review of water pollution control. Stresses the legal and technological attempts at control.

McGauhey, P.H. "Man-Made Contamination Hazards to Ground Water." GROUND WATER 6 (May-June 1968): 10-13.

General discussion of the different processes by which the ground water is being contaminated and the danger associated with this contamination. Contamination from each of the following sources is discussed: human wastes, industrial processes, agricultural irrigation water, and solid residues from industry and mining.

Mackenthun, Kenneth M. THE PRACTICE OF WATER POLLUTION BIOLOGY. Washington, D.C.: U.S. Department of Interior, Federal Water Pollution Control Administration, 1969. 281 p.

Good beginning introduction to water pollution with emphasis on the biological aspects. Covers water quality constituents, how to collect and analyze data, plus the effects of various pollutants on streams.

______. TOWARD A CLEANER AQUATIC ENVIRONMENT. Washington, D.C.: Environmental Protection Agency, Office of Air and Water Programs, 1973. 273 p.

Introduction to the problems of water pollution, what it is, controls on pollutors, and how the environment responds to pollution. Describes the techniques used to investigate water pollution and how to interpret the results.

Mackenthun, Kenneth M., and Ingram, William M. BIOLOGICAL ASSOCIATED PROBLEMS IN FRESHWATER ENVIRONMENTS: THEIR IDENTIFICATION, INVESTIGATION AND CONTROL. Washington, D.C.: U.S. Department of Interior, Federal Water Pollution Control Administration, 1967. x, 287 p. Glossary.

General survey of the aquatic environment and problems associated with escessive plant growth, aquatic animal pests, and slimes. Indicates methods of control.

Marx, Wesley. THE FRAIL OCEAN. New York: Coward-McCann, 1967. viii, 248 p.

Popular text, examining the effects of man on the oceans primarily recounting the most recent abuses in the manner of Rachel Carson.

Mellanby, K[enneth]. THE BIOLOGY OF POLLUTION. Studies in Biology, no. 38. London: Edward Arnold Publishers, 1972. 59 p.

Provides a very general overview and introduction to pollution for those with little or no previous background.

Miller, Stanton S., ed. WATER POLLUTION. Washington, D.C.: American Chemical Society, 1974. 317 p.

Collection of 106 articles from ENVIRONMENTAL SCIENCE AND TECHNOLOGY surveying current problems and present practices in water pollution. Articles cover monitoring, specific pollutants, wastewater treatment, desalination, adsorption, microstaining, ocean dumping, and tertiary treatment.

Montgomery, Austin H., Jr. "Water Pollution Control: A History." WATER AND SEWAGE WORKS 119 (February 1972): 60-65.

Review of man's attempts to adequately dispose of his own wastes.

Murphy, R. Sage, and Nyquist, David, eds. INTERNATIONAL SYMPOSIUM ON WATER POLLUTION CONTROL IN COLD CLIMATES. Washington, D.C.: Environmental Protection Agency, Research and Monitoring, 1970. U.S. Government Printing Office. x, 332 p.

Collection of technical papers presented at a symposium relating to: the effects of pollutants on receiving waters in cold climates, and the treatment of pollutants in cold climates. Studies are applicable to other climates.

Oglesby, Ray T.; Carlson, Clarence A.; and McCann, James A., eds. RIVER ECOLOGY AND MAN. New York: Academic Press, 1972. xvii, 465 p.

Collection of general review articles given at a symposium. The text is divided into four sections: river ecology, general effects of man on river systems, specific effects of man on rivers such as thermal pollution and the addition of nutrients, and the use of river, for example, economic.

Pettyjohn, Wayne A., ed. WATER QUALITY IN A STRESSED ENVIRONMENT: READINGS IN ENVIRONMENTAL HYDROLOGY. Minneapolis: Burgess Publishing, 1972. ix, 309 p.

Collection of semitechnical to popular articles relating to water pollution, many on ground water pollution. Problems of water pollution are presented mainly as case studies and involve such topics as the legal aspects, and trace elements in drinking water.

Polunin, Nicholas, ed. THE ENVIRONMENTAL FUTURE: PROCEEDINGS OF THE FIRST INTERNATIONAL CONFERENCE ON ENVIRONMENTAL FUTURE. New York: Barnes and Noble, 1972. xiv, 660 p.

> General discussion articles on the various aspects of man's effect on nature including water. Three of the chapters are concerned directly with water.

Rabin, Edward H., and Schwartz, Mortimer D., eds. THE POLLUTION CRISIS: OFFICIAL DOCUMENTS. Dobbs Ferry, N.Y.: Oceana Publications, 1972. 510 p.

> Collection of official government documents relating to the various aspects of pollution. Section on water pollution includes "Clean Water for the 1970's, A Status Report," "Ocean Dumping, A National Policy," "National Estuary Study," and others.

Renn, Charles E. OUR ENVIRONMENT BATTLES WATER POLLUTION. Chestertown, Md.: LaMotte Chemical Products, 1969. 32 p.

> Short elementary text on water pollution covering such areas as the nitrogen and carbon dioxide cycle, the hydrologic cycle, the effect of pollutants on aquatic systems, aquatic pests, and water purification. Presents a general overview of the subject.

Ruivo, Mario, ed. MARINE POLLUTION AND SEA LIFE. F.A.O. TECHNICAL CONFERENCE ON MARINE POLLUTION AND ITS EFFECT ON LIVING RESOURCES AND FISHING. West Byfleet, Engl.: Fishing News (for) the Food and Agricultural Organization of the United Nations, 1972. xxiv, 624 p.

> Conference covered following areas: marine pollution in the world today, behavior and fate of pollutants, effect of pollutants on biology and life cycles, ecosystem modification and effect on marine communities, technical aspects of minimizing pollution and countering its effects, effect of pollutants on quality of marine products and effects on fishing, and legislature control.

Sibthorp, M.M. OCEANIC POLLUTION: A SURVEY AND SOME SUGGESTIONS FOR CONTROL. London: David Davies Memorial Institute of International Studies, 1969. 53 p.

> Deals very generally with oceanic pollution indicating sources of pollution and problems of control. Lists international organizations concerned with pollution, as well as national bodies concerned with oceanography.

Tarzwell, C.M., ed. BIOLOGICAL PROBLEMS IN WATER POLLUTION. Public Health Service Publication, no. 999-WP-25. Cincinnati: U.S. Public Health Service, 1965. ix, 424 p.

Contains a collection of semitechnical review articles relating to water quality, pollution effects and environment requirements of algae, freshwater invertebrates, marine invertebrates, aquatic insects, fish, and wildlife. Includes effects of pesticides, metal toxicants, and radionuclides on biological systems.

U.S. Council on Environmental Quality. ENVIRONMENTAL QUALITY. Washington, D.C.: U.S. Government Printing Office, 1970-- . Annual.

Each report indicates, in general: where we're at, the costs involved in pollution control during the previous year, the significant laws and decisions. Indicates what pollutants were in the forefront of attention, and also gives general statistics for expenditures, water uses, discharges, and spills.

U.S. Environmental Protection Agency. NATIONAL GROUND WATER QUALITY SYMPOSIUM. Water Pollution Control Research Series 16060 GRB 08/71. Washington, D.C.: U.S. Government Printing Office, 1972. iii, 211 p.

Series of semitechnical to technical papers on ground water pollution, many of which cite case studies of pollution, while others discuss the effects of disposal of wastes under ground and evaluate the problem and methods of clean up.

Vernberg, F.J., and Vernberg, W.B., eds. POLLUTION AND PHYSIOLOGY OF MARINE ORGANISMS. New York: Academic Press, 1974. xiii, 492 p.

Collection of papers which deals with the effects of pollutants on marine organisms. Book is divided into six sections: heavy metals, pesticides, PCBs and oil, general papers, factors interacting with pollutants, and other environmental parameters. Papers range in scope from general to detailed and highly technical, with the majority in the latter category.

Warren, Charles E. BIOLOGY AND WATER POLLUTION CONTROL. Philadelphia: W.B. Saunders, 1971. xiv, 434 p.

Title of this book is somewhat misleading as it deals at great length with aquatic biology. Includes chapters on the historical aspects and meaning of pollution, water quality standards, toxicity, and tolerance conditions and waste treatment, but its greatest contribution is in the excellent discussions on aquatic ecology.

Wilber, Charles G. THE BIOLOGICAL ASPECTS OF WATER POLLUTION. Springfield, Ill.: Charles C Thomas, 1969. vii, 296 p.

Text on the biological effects of many water pollutants such as metals, oil, pesticides, pulp mill pollution, and sanitary sewage. Cites background information on toxicology, ecology, and general water chemistry.

Chapter 15
WATER POLLUTANTS

A. PESTICIDES

The variety of pesticides and their usage has increased tremendously in the past ten to fifteen years. While DDT predominated the earlier literature, "alphabet soup" is an apt description for today's pesticide literature. The focus of research has also changed. Where earlier pesticide studies were concerned only with the pure, laboratory-produced compound, it is now recognized that pesticides are chemically transformed either through interaction with other chemicals in the application media, or by the environment after application. Recent studies (see section on journal articles, p. 89) focus on the manner of transformation, the detection and distribution of the transformed products, and their toxicity. In addition, the widespread distribution of pesticide residues within aquatic environments, especially their concentration in the food chain, has resulted in more integrative ecosystem studies (see Bruce et al., p. 89, and Isensee and Jones, p. 89) on pesticide movement.

With the continued production of new pesticides, the reader should check the abstracts and journals (see "Journals and Newsletters" under section D, Pesticide and Toxicology, p. 49) for the most recent literature. Additional references to pesticides may be found in "Pollution Sources" under section B, Agricultural Pollution, p. 139.

1. Abstracts and Bibliographies

Diaz-Colon, J.D., and Bovey, R.W. PESTICIDES IN WATER: A LIST OF SELECTED REFERENCES. College Station: Texas A & M University, Agricultural Experiment Station, 1973. 53 p. Paperbound.

> Bibliographic citations of references which report the fate, persistence, and mobility of pesticides in water, their effects on aquatic life, and analytical procedures.

Rudd, Robert L. ENVIRONMENTAL TOXICOLOGY: A GUIDE TO INFORMATION SOURCES. Man and the Environment Series, vol. 7. Detroit: Gale Research Co., 1977.

Annotated bibliography of sources of information on environmental toxicology, the effects of environmental pollution, and special aspects of the problem including legislation and regulation, monitoring, animal resistance to pollutants, and alternatives to chemical control of pests. Annotations are arranged by category of pollutant. Annotations include many articles on effects of toxic substances on aquatic systems and aquatic organisms, and the subject index permits location of annotations concerned with effects of any particular pollutant.

[U.S. Department of Interior.] Water Resources Scientific Information Center. ALDRIN AND ENDRIN IN WATER: A BIBLIOGRAPHY. Bibliography Series WRSIC 72-203. Washington, D.C.: 1972. Distributed by National Technical Information Service. 120 p.

Annotated bibliography of articles dealing with aldrin and endrin in freshwater. Abstracts are from SELECTED WATER RESOURCES ABSTRACTS, and cover the period up to 15 February 1972.

_____. DDT IN WATER: A BIBLIOGRAPHY. Bibliography Series, WRSIC 71-211. Washington, D.C.: 1971. Distributed by National Technical Information Service. 278 p.

Collection of abstracts from SELECTED WATER RESOURCES ABSTRACTS (up to 15 October 1971) on all aspects of DDT in water.

_____. DIELDRIN IN WATER: A BIBLIOGRAPHY. Bibliography Series, WRSIC 72-202. Washington, D.C.: 1972. Distributed by National Technical Information Service. 144 p.

Abstracts from SELECTED WATER RESOURCES ABSTRACTS (up to 15 December 1971) which relate to dieldrin in water.

2. References

American Chemical Society. FATE OF ORGANIC PESTICIDES IN THE AQUATIC ENVIRONMENT. Advances in Chemistry Series, III. Washington, D.C.: 1972. viii, 280 p.

Papers presented at the sixty-first meeting of the American Chemical Society (March 1971) on the physical and chemical processes affecting pesticides in water, as well as analytical procedures.

Bloom, Sandra C., and Degler, Stanley E. PESTICIDES AND POLLUTION. Washington, D.C.: Bureau of National Affairs, 1969. iii, 99 p.

Text deals with the dangers that pesticides pose to the environment, the interests that industry has in the production of pesticides, and the role of government agencies in controlling pesticide use.

Cornell Aeronautical Laboratory. THE MOVEMENT AND IMPACT OF PESTICIDES USED IN FOREST MANAGEMENT ON THE AQUATIC ENVIRONMENT AND ECOSYSTEM. Pesticide Study Series, 7. Washington, D.C.: Environmental Protection Agency, Office of Water Programs, 1972. Various pagination.

Report relates to the use and application of pesticides in the forests of New York. Includes general information on the routes of pesticide entry into aquatic environments and their effect on aquatic organisms. Also includes laws and regulations of New York pertaining to the use of pesticides.

Edwards, Clive A. PERSISTENT PESTICIDES IN THE ENVIRONMENT. 2d ed. Cleveland: CRC Press, 1973. 170 p.

Review of the occurrence of pesticide residues in the environment, including water and in the biota. Includes many statistics indicating pesticide levels in soils, air, rainwater, water, soil flora and fauna, aquatic organisms, plants, fish, man, and other vertebrates.

______, ed. ENVIRONMENTAL POLLUTION BY PESTICIDES. Environmental Science Research, vol. 3. New York: Plenum Press, 1973. x, 542 p.

Book is a compilation of papers which review the literature concerning pesticide residues in the environment, their dynamics, degradation, effect on organisms, and analysis techniques. Contents include pesticide residues in fish, effect of pesticides on fish, and residues in water.

Evans, John O., and Duseja, D.R. HERBICIDE CONTAMINATION OF SURFACE RUNOFF WATERS. Environmental Protection Technical Series. Washington D.C.: Office of Research and Monitoring, U.S. Environmental Protection Agency, 1973. Distributed by U.S. Government Printing Office. x, 99 p.

Report of a field and laboratory study on the movement of herbicides, as contaminants of irrigation return flow. The primary emphasis of the study was on herbicides used for vegetation control along ditches and canals.

Food Protection and Toxicology Center. University of California at Davis. THE EFFECTS OF AGRICULTURAL PESTICIDES IN THE AQUATIC ENVIRONMENT, IRRIGATED CROPLANDS, SAN JOAQUIN VALLEY. Pesticide Study Series, 6. Washington, D.C.: Environmental Protection Agency, Office of Water Programs, 1972. iii, S22, 268 p.

Overview of the effects of pesticides on the aquatic environment with information pertinent to the San Joaquin Valley,

California. Includes an inventory of pesticide uses, application techniques, route of entry in water, degradation, metabolism, and persistence in water, as well as impact of pesticide pollution on the aquatic environment and the laws and regulations governing pesticide usage in the valley. Most of the information is of a general nature with a good literature review.

Gillett, James W., ed. THE BIOLOGICAL IMPACT OF PESTICIDES IN THE ENVIRONMENT: A SYMPOSIUM ASSESSING THE SIGNIFICANCE OF PESTICIDES IN RELATION TO ECOLOGICAL PROBLEMS AND HEALTH. Corvallis: Environmental Health Sciences Center, Oregon State University, 1970. Distributed by Oregon State University Press. xi, 210 p.

Compilation of papers given at a symposium concerned primarily with the effects of pesticides on organisms. Fish were the only aquatic organism discussed. Includes general discussions on the transport and accumulation of pesticides.

Matsumura, Fumio. TOXICITY OF INSECTICIDES. New York: Plenum Publishing, 1975. 504 p.

Survey of the toxicology of insecticides including insecticide chemistry, mechanisms of action, and metabolism. The book reviews topics such as metabolism in animals, plants and microorganisms, the dynamics of insecticide movement in animals and environment, and the effects of pesticides on wildlife. Book will be of interest to chemists, engineers, zoologists, physiologists, and environmental scientists.

Matsumura, Fumio; Boush, G. Mallory; and Misato, Tomomasa, eds. ENVIRONMENTAL TOXICOLOGY OF PESTICIDES. New York: Academic Press, 1972. 637 p.

Report on the present state of knowledge about pesticide contamination of the environment. Subjects covered are the pattern of pesticide usage, effect of environmental factors, both physical and biological, on pesticide degradation, and environmental contamination by fungicides, herbicides, and organophosphates.

Mellanby, Kenneth. PESTICIDES AND POLLUTION. 2d rev. ed. London: Collins, 1970. 221 p.

Presents a brief overview of air and water pollution in the first four chapters, with the remaining six examining the use and misuse of pesticides. For each pesticide cited, information is given on its use and effectiveness, and on its effect on nontarget species, both plant and animal. Most examples are British.

Midwest Research Institute, Kansas City, Mo. THE POLLUTION POTENTIAL IN PESTICIDE MANUFACTURING. Pesticide Study Series, 5. Washington, D.C.: Environmental Protection Agency, Office of Water Programs, 1972. viii, 249 p.

Survey and evaluation of the environmental pollution potential associated with the manufacture, formulation, and marketing of pesticides. Includes data on present (1972) production, sites, manufacturing methods, and detailed information on twenty-two pesticides, following each step of their manufacturing and marketing process.

Muirhead-Thomson, R.C. "Laboratory Evaluation of Pesticide Impact on Stream Invertebrates." FRESHWATER BIOLOGY 3 (October 1973): 479-98.

Description of laboratory techniques for evaluating the reaction, i.e., mortality, of representative stream invertebrates to known time/concentrations of a variety of pesticides, including DDT, toxaphene, parathion, carbaryl, dieldrin, and endosulphan. Conditions for the experiments included continuous water flow in test containers and exposure to the commercial pesticide formulation instead of the pure chemical to simulate the natural situation. Implications of the results are discussed with particular reference to pesticide impact on predator/prey relationships.

______. PESTICIDES AND FRESHWATER FAUNA. New York: Academic Press, 1971. vi, 248 p.

Book is primarily concerned with the use of pesticides to control trash fish and disease or pest organisms. Includes search for selective pesticides, and studies on toxicity and effectiveness of these pesticides. Two chapters cover toxicity to nontarget organisms and the effects of physical and chemical factors on toxicity.

National Research Council. CHLORINATED HYDROCARBONS IN THE MARINE ENVIRONMENT, A REPORT. Washington, D.C.: National Academy of Sciences, 1971. vii, 42 p.

Review of the effects and accumulation of DDT and related compounds in the oceans. Includes recommendations for studies to determine the effects of these chemicals on the community structure of marine ecosystems and to develop more base line data.

Pimentel, David. ECOLOGICAL EFFECTS OF PESTICIDES ON NON-TARGET SPECIES. Washington, D.C.: U.S. Government Printing Office, 1971. 220 p.

Comprehensive compilation of published data on the environmental effects of pesticides. For each pesticide, information is presented about its influence on nontarget flora and fauna, as well as its concentration in food chains and persistence in the environment.

Rumker, R.V., et al. THE USE OF PESTICIDES IN SUBURBAN HOMES AND GARDENS AND THEIR IMPACT ON THE AQUATIC ENVIRONMENT. Pesticide Study Series, 2. Washington, D.C.: Environmental Protection Agency, Office of Water Programs, 1972. Distributed by U.S. Government Printing Office. Various pagination.

> Evaluation of the impact on natural environments of pesticides used in the suburban homes and gardens of Philadelphia, Lansing and Dallas. Presents known information on pesticide use, transport mechanisms, degradation, aquatic impact, alternatives, and laws and regulations governing the use of pesticides.

Smithsonian Institution. Science Information Exchange. A CATALOG OF RESEARCH IN AQUATIC PEST CONTROL AND PESTICIDE RESIDUES IN AQUATIC ENVIRONMENTS. Pesticide Study Series, 1. Washington, D.C.: Environmental Protection Agency, Office of Water Programs, 1972. Distributed by U.S. Government Printing Office. Various pagination.

> Compilation of research projects related to the subjects in the title, indicates objectives, approach, progress, and researchers.

Teledyne Brown Engineering. PESTICIDE USAGE AND ITS IMPACT ON THE AQUATIC ENVIRONMENT IN THE SOUTHEAST. Pesticide Study Series, 8. Washington, D.C.: Environmental Protection Agency, Office of Water Programs, 1972. vi, 411 p.

> Critical examination of pesticide usage and its impact on the aquatic environment of the southeast. Includes data on pesticide usage and its effects, summaries of current regulatory statutes and actual case studies with literature citations.

U.S. Agricultural Research Service. Plant Pest Control Division. MONITORING AGRICULTURAL PESTICIDE RESIDUES: A PRELIMINARY REPORT OF STUDIES ON SOIL, SEDIMENT AND WATER IN MISSISSIPPI RIVER DELTA. Washington, D.C.: U.S. Department of Agriculture, 1966. 53 p.

> Results of a study which analyzed pesticide levels at various points within the Mississippi River Delta. Extensive data presented from the study which was conducted in 1964.

U.S. Environmental Protection Agency. EFFECTS OF PESTICIDES IN WATER: A REPORT TO THE STATES. Washington, D.C.: 1973. v, 146 p.

> Report consists primarily of tables which list pesticide tested, organism tested and the toxic concentration. Includes information on the nature of the environmental effect of the pesticide and the economic and social impact of control measures. Good source of statistics.

U.S. Environmental Protection Agency. Office of Water Programs. PESTICIDES IN THE AQUATIC ENVIRONMENT. Washington, D.C.: 1972. 181 p.

Report on the environmental damage done by pesticides in the aquatic environment; consists of current information on the effects of pesticides, as well as the nature of these effects, and the impact and practicality of control measures.

3. Journal Articles

Bruce, R.R.; Harper, L.A.; Leonard, R.A.; Snyder, W.M.; and Thomas, A.W. "A Model for Runoff of Pesticides from Small Upland Watersheds." JOURNAL OF ENVIRONMENTAL QUALITY 4 (October and December 1975): 541-48.

Earnest, R.D., and Benville, P.E., Jr. "Correlation of DDT and Lipid Levels for Certain San Francisco Bay Fish." PESTICIDE MONITORING JOURNAL 5 (December 1971): 235-41.

Foehrenbach, J.; Mahmood, G.; and Sullivan, D. "Chlorinated Hydrocarbon Residues in Shellfish (Pelecypoda) from Estuaries of Long Island, New York." PESTICIDE MONITORING JOURNAL 5 (December 1971): 242-47.

Gilluly, R.H. "Oceans as Alphabet Soup: Focus on DDT and PCB's." SCIENCE NEWS 101 (8 January 1972): 30-31.

Harvey, G.R.; Miklas, H.P.; Bowen, V.T.; and Steinhauer, W.G. "Observations on the Distribution of Chlorinated Hydrocarbons in Atlantic Ocean Organisms." JOURNAL OF MARINE RESEARCH 32 (May 1974): 103-18.

Holden, V.A. "The Effect of Pesticides on Life in Fresh Waters." PROCEEDINGS OF THE ROYAL SOCIETY OF LONDON, Series B, 180 (March 1972): 383-94.

Isensee, A.R., and Jones, G.E. "Distribution of 2, 3, 7, 8-Tetrachlorodiobenzo-P-dioxin (TCDD) in Aquatic Model Ecosystems." ENVIRONMENTAL SCIENCE AND TECHNOLOGY 9 (July 1975): 668-72.

Leland, H.V.; Bruce, W.N.; and Shimp, N.F. "Chlorinated Hydrocarbon Insecticides in Sediments of Southern Lake Michigan." ENVIRONMENTAL SCIENCE AND TECHNOLOGY 7 (September 1973): 833-38.

Maddox, J. "Pollution and Worldwide Catastrophe." NATURE 236 (28 April 1972): 433-36.

Miles, J.R.W., and Harris, C.R. "Insecticide Residues in a Stream and a Controlled Drainage System in Agricultural Areas of Southwest Ontario, 1970." PESTICIDE MONITORING JOURNAL 5 (December 1971): 289-94.

Moriarty, F. "Pollutants and Food Chains." NEW SCIENTIST 56 (16 March 1972): 594-96.

Robertson, L.W., and Chynoweth, D.P. "Another Hallogenated Hydrocarbon." ENVIRONMENTAL SCIENCE AND TECHNOLOGY 10 (January 1976): 25-27.

Sanders, H.O., and Chandler, J.H. "Biological Magnification of a Polychlorinated Biphenyl (Aroclor 1254) from Water by Aquatic Invertebrates." BULLETIN OF ENVIRONMENTAL CONTAMINATION AND TOXICOLOGY 7 (May 1972): 257.

Wallace, J.B., and Brady, U.E. "Residue Levels of Dieldrin in Aquatic Invertebrates and Effect of Prolonged Exposure on Populations." PESTICIDE MONITORING JOURNAL 5 (December 1971): 295-300.

Walsh, F., and Mitchell, R. "Inhibition of Inter-microbial Predation by Chlorinated Hydrocarbons." NATURE 249 (14 June 1974): 673-74.

Woodwell, G.M.; Craig, P.P.; and Johnson, H.A. "DDT in the Biosphere: Where Does It Go." SCIENCE 174 (10 December 1971): 1101-08.

B. DETERGENTS

In the 1960s the most noticeable feature of detergent pollution was massive accumulations of foam. Besides causing visual pollution, the nonbiodegradable detergent was toxic to many aquatic organisms including fish.

In 1965 the detergent industry switched from ABS (alkylbenzene sulfonate) detergents to biodegradable LAS (linear alkylsulfonate) detergents. While detergents now produce little foam, they are still a major source of phosphorus which is considered a significant contributor to eutrophication. The reader should consult chapter 15, section I, p. 125, on eutrophication and organic pollutants for additional references on phosphorus pollution. References on the toxicity of specific detergent components and of proposed new compounds have been included in the journal articles section, p. 92.

1. Abstracts and Bibliographies

Evans, Francis, L.; Cohen, Jesse M.; and Hannah, Sidney A., eds. BIBLIOGRAPHY ON SYNTHETIC DETERGENTS IN WATER AND WASTES: INCLUDING ANALYTICAL METHODS AND PHYSIOLOGICAL EFFECTS. Cincinnati: U.S. Department of Health, Education and Welfare, Public Health Service, Robert A. Taft Sanitary Engineering Center, 1964. 91 p.

> Contains over 850 references from 1937 through June 1964 listed by year and alphabetical order within each year. No index.

[U.S. Department of Interior.] Water Resources Scientific Information Center. DETERGENTS IN WATER: A BIBLIOGRAPHY. Bibliography Series WRSIC 71-214. Washington, D.C.: 1971. Distributed by National Technical Information Service. 460 p.

Citations from SELECTED WATER RESOURCES ABSTRACTS, 1966 through 1971, and NOTICES OF RESEARCH PROJECTS.

2. References

Battelle Memorial Institute. Columbus Laboratories. CORROSION POTENTIAL OF NTA IN DETERGENT FORMULATIONS. Washington, D.C.: Water Quality Office, U.S. Environmental Protection Agency, 1971. x, 94 p.

Study of the effect of NTA on the machinery employed in laundering and in a sewer system, in addition to its chelation properties.

California State Water Quality Control Board. DETERGENT REPORT: A STUDY OF DETERGENTS IN CALIFORNIA. Sacramento: California State Legislature, 1965. ix, 55 p.

Report on the use of detergents, their chemistry and toxicity, as well as their occurrence in the waters of California. Provides an historical background on the problems associated with detergents, particularly the ABS type.

Dunlap, W.J., et al. INVESTIGATIONS CONCERNING PROBABLE IMPACT OF NITRILOTRIACETIC ACID ON GROUND WATERS. Water Pollution Control Research Series 16060 GHR 11/71. Washington, D.C.: U.S. Environmental Protection Agency, 1971. ix, 51 p.

Discussion of the fate and impact of nitrilotriacetic acid in different soil types and in ground water.

Howe, Richard S., et al. LAUNDRY DETERGENTS AND ENVIRONMENTAL QUALITY. Occasional Papers, no. 2. Bloomington: University of Indiana, School of Public and Environmental Affairs, 1973. 24 p. Pamphlet.

Evaluates the feasibility of reducing phosphorus content of detergents, i.e., the cost of reduction and the subsequent cleaning ability of the detergent. Includes effect of detergents on water supply.

I.T.T. Research Institute. Technology Center. DEVELOPMENT OF PHOSPHATE-FREE HOME LAUNDRY DETERGENTS. Water Pollution Control Research Series 16080 DVF 12/70. Washington, D.C.: U.S. Environmental Protection Agency, 1970. x, 103 p.

Report of a study on formulation of new phosphate-free detergents and an evaluation of their ability to stimulate algal growth.

Prat, Jean, and Giraud, A. THE POLLUTION OF WATER BY DETERGENTS. Paris: Organization for Economic Cooperation and Development, 1964. 86 p.

Technical report on the use, toxicity, and determination of detergents, indicating their chemistry. Good background information.

Thom, N.S. "Nitrilotriacetic Acid: A Literature Survey." WATER RESEARCH 5 (July 1971): 391-99.

Literature review of NTA, discusses its biodegradability, its effects on sludge and on freshwater organisms.

3. Journal Articles

Angino, E.E.; Magnuson, L.M.; Waugh, T.C.; Galle, O.K.; and Bredfeldt, J. "Arsenic in Detergents: Possible Danger and Pollution Hazard." SCIENCE 168 (17 April 1970): 389-90.

Biesinger, K.E.; Andrew, R.W.; and Arthur, J.W. "Chronic Toxicity of NTA (nitrolotriacetate) and Metal-NTA Complexes to Daphnia Magna." JOURNAL FISHERIES RESEARCH BOARD OF CANADA 31 (April 1974): 486-90.

Cook, T.M., and Goldman, C.K. "Degradation of Anionic Detergents in Chesapeake Bay." CHESAPEAKE SCIENCE 15 (March 1974): 52-55.

Doemel, W.N., and Brooks, A.E. "Detergent Phosphorus and Algal Growth." WATER RESEARCH 9 (August 1975): 713-19.

Emden, H.M., van; Kroon, C.C.M.; Schoeman, E.N.; and Seventer, H.A., van. "The Toxicity of Some Detergents Tested on Aedes Aegypti L., Lebistes Reticulatus Peters and Biomphalaria Glagrata (Say)." ENVIRONMENTAL POLLUTION 6 (June 1974): 297-308.

Gutenmann, W.H., and Lisk, D.J. "Flame Retardant Release from Fabrics during Laundering and Their Toxicity to Fish." BULLETIN OF ENVIRONMENTAL CONTAMINATION AND TOXICOLOGY 14 (July 1975): 61-64.

Macek, K.J., and Krzeminski, S.F. "Susceptibility of Bluegill Sunfish (Lepomis Macrochirus) to Nonionic Surfactants." BULLETIN OF ENVIRONMENTAL CONTAMINATION AND TOXICOLOGY 13 (March 1975): 377-84.

McKim, J.M.; Arthur, J.W.; and Thorslund, T.W. "Toxicity of a Linear Alkylate Sulfonate Detergent to Larvae of Four Species of Freshwater Fish." BULLETIN OF ENVIRONMENTAL CONTAMINATION AND TOXICOLOGY 14 (July 1975): 1-7.

Pattison, E.S. "Arsenic and Water Pollution Hazard." SCIENCE 170 (20 November 1970): 870.

Porcella, D.B.; Cowan, P.A.; and Middlebrooks, E.J. "Biological Response to Detergent and Non-detergent Phosphorus in Sewage. Part II." WATER AND SEWAGE WORKS INCLUDING INDUSTRIAL WASTES 120 (December 1973): 36-37.

Silberman, H.C. "Effects of Carbonate Detergent on Biological Wastewater Treatment." JOURNAL WATER POLLUTION CONTROL FEDERATION 47 (March 1975): 524-34.

Sturm, R.N.; Williams, K.E.; and Macek, K.J. "Fluorescent Whitening Agents: Acute Fish Toxicity and Accumulation Studies." WATER RESEARCH 9 (February 1975): 211-19.

Tobin, R.S.; Onuska, F.I.; Anthony, D.H.J.; and Comba, M.E. "Nonionic Surfactants: Conventional Biodegradation Test Methods Do Not Detect Persistent Polyglycol Products." AMBIO 5, no. 1 (1976): 30-31.

Yentsch, C.M.; Yentsch, C.S.; Owen, C.; and Salvaggio, M. "Stimulatory Effects on Growth and Photosynthesis of the Toxic Red Tide Dinoflagellate, Gonyaulax Tamarensis with the Addition of Nitrolotriacetic Acid (NTA) $N(CH_2COOH)_3$." ENVIRONMENTAL LETTERS 6, no. 2 (1974): 231-38.

C. OIL

The increasing use of oil, both as an energy source and as an ingredient in many manufactured products, has resulted in a worldwide pollution problem. Even uninhabited areas are susceptible to oil pollution from oil tanker accidents or oil pipeline leaks. While most people are aware of these dramatic and highly publicized sources of oil pollution, less noticeable but no less destructive sources are found closer to home. These include oily wastes from industry (see section on industrial pollution, p. 133, for additional references), refineries and from the storage and distribution of oil. Problems also arise from the disposal of oil from garages, street runoff, and pleasure boats. All these sources have contributed to the ubiquitousness of oil today.

Research on oil pollution is multifaceted; there are the legal and administrative aspects involving jurisdiction and responsibility for cleanup, the regulation of oil tankers, and the control of waste discharges from ships and industry. Other areas of research include the detection and removal of oil, the determination of its effects on aquatic organisms, and the degradation of oil by the living and nonliving components of the environment.

Water Pollutants

The following references provide an introduction to all these research areas. The journal literature generally contains more recent information and is indicative of the journals having oil pollution literature.

1. Abstracts and Bibliographies

Bates, Martha. WATER POLLUTION BY OIL SPILLAGE. 2 vols. Dallas: Southern Methodist University, Industrial Information Services, Science Library, 1969.

> Volumes consist primarily of abstracts of the literature between 1964-69 on the causes, effects, and methods of alleviating oil spills. Includes legal and economic aspects.

Christol, Carl Q. OIL POLLUTION OF THE MARINE ENVIRONMENT. Serial 92-1. Washington, D.C.: U.S. Government Printing Office, 1971. ix, 93 p. Paper.

> Legal bibliography concerned with the functions of nations, international organizations and the law with regard to oil pollution. Emphasizes legal rights and duties.

Holmes, Robert W., ed. OIL POLLUTION: AN INDEX-CATALOG TO THE COLLECTION OF THE OIL SPILL INFORMATION CENTER. 4 vols. Santa Barbara: University of California Library, 1972.

> Index of oil pollution information for the period 1969-71 held by the Library. Includes information from the Torrey Canyon incident (1967). Author, title, and source are given for each reference, as well as location in the University of California at Santa Barbara Library.

Moulder, David S., and Varley, Allen, eds. A BIBLIOGRAPHY ON MARINE AND ESTUARINE OIL POLLUTION. Devon, Engl.: Laboratory of the Marine Biological Association of the United Kingdom, 1971. vii, 129 p.

> References about one thousand papers on marine and estuarine oil pollution covering the past one hundred years up to and including 1971. Nonscientific or journalistic articles not included.

Radcliffe, Donna R., and Murphy, Thomas A. BIOLOGICAL EFFECTS OF OIL POLLUTION: BIBLIOGRAPHY. Water Pollution Control Series. Washington, D.C.: Federal Water Pollution Control Administration, 1969. Distributed by National Technical Information Service, U.S. Department of Commerce, Springfield, Va. 46 p.

> Collection of references for the period up to mid-1969 on general aspects of oil pollution, as well as reports of oil spills, general biological effects, and effects on specific organisms such as birds, fish, shellfish, plants, and invertebrates.

[U.S. Department of Interior.] Water Resources Scientific Information Center. OIL SPILLAGE: A BIBLIOGRAPHY. 2 vols. Washington, D.C., 1973.

Abstracts from SELECTED WATER RESOURCES ABSTRACTS up to February 1973 covering all aspects of oil pollution, including patents for oil spill removal apparatus.

Walstad, Kay, and Hearth, Fred E., eds. OIL POLLUTION IN THE SANTA BARBARA CHANNEL: A COMPREHENSIVE BIBLIOGRAPHY. Santa Barbara: University Library, University of California, 1972. vii, 195 p. Photos.

Survey of the Santa Barbara oil spill covers the period 1969 through 1971. Includes all articles, even newspaper accounts, and motion pictures.

2. References

American Petroleum Institute. PREVENTION AND CONTROL OF OIL SPILLS. New York: 1970. x, 345 p.

Proceedings of a conference which covered a wide range of topics from the administration to the physical chemistry involved in oil dispersal. Articles are generally of a discussion type and range from technical to nontechnical.

Blumer, Max. "Oil Pollution of the Ocean." In MAN'S IMPACT ON ENVIRONMENT, edited by Thomas R. Detwyler, pp. 295-301. New York: McGraw-Hill, 1971.

Brief overview of the physical extent and biological effects of oil spills and of the present countermeasures used in oil spill cleanups.

California. University of. Santa Barbara. SANTA BARBARA OIL POLLUTION 1969: A STUDY OF BIOLOGICAL EFFECTS OF THE OIL SPILL. Washington, D.C.: Federal Water Quality Administration, Department of Interior, 1970. 49 p.

Report on the effects of the Santa Barbara oil spill on intertidal plants and animals, plus kelp beds.

______. SANTA BARBARA OIL SYMPOSIUM: OFFSHORE PETROLEUM PRODUCTION; AN ENVIRONMENTAL INQUIRY. Washington, D.C.: U.S. Government Printing Office, 1972. 377 p.

The Santa Barbara oil spill as perceived by social, economic, industrial, and environmental spokesmen. Speakers included the head of Humble Oil Geological Survey, as well as faculty and students of the university and residents of the city of Santa Barbara.

California State Water Resources Control Board. EVALUATING OIL SPILL CLEANUP AGENTS. Publication no. 43 of the Board. Sacramento, Calif.: 1971. 150 p.

Results of a study which evaluated the chemical and biological effects of various oil spill cleanup agents. Contents include a review of the effect of oils and chemical dispersants on aquatic life, results, and evaluation of the toxicity of dispersants, the physical chemistry of dispersants in water, and performance effectiveness, including degradability.

Carthy, John D., and Arthur, Don R., eds. SYMPOSIUM ON THE BIOLOGICAL EFFECTS OF OIL POLLUTION ON LITTORAL COMMUNITIES. London: Field Studies Council, 1968. 197 p.

Papers discuss the chemistry of oil and of the detergents used in the cleanup, modes of removal and effects of oil on various populations in the littoral community. Some discussion on the use of detergents for oil cleanup. Lengthy bibliography on oil pollution is included.

Chan, Gordon L. A STUDY OF THE EFFECTS OF THE SAN FRANCISCO OIL-SPILL ON MARINE LIFE. 2 parts. Kentfield, Calif.: College of Marin, 1972.

Report on the analysis of the Duxbury Reef oil spill including samplings and statistical procedures, as well as charts and tables of the results.

Cowell, E.B., ed. ECOLOGICAL EFFECTS OF OIL POLLUTION ON LITTORAL COMMUNITIES. London: Institute of Petroleum, 1971. viii, 250 p.

Topics include effect of oil on the biological communities and the effects of cleanup agents.

Griner, Lynn A., and Herdman, Robert. EFFECTS OF OIL POLLUTION ON WATERFOWL--A STUDY OF SALVAGE METHODS. Water Pollution Control Research Series. Washington, D.C.: U.S. Environmental Protection Agency, Water Quality Office, 1970. ii, 35 p.

Report of a study of salvage methods for waterfowl, primarily mallard ducks. Includes pathology and behavioral alterations noted in the ducks. Presents recommendations for care and cleaning agents.

Hepple, Peter, ed. WATER POLLUTION BY OIL. New York: Elsevier Publishing, 1971. viii, 393 p.

Collection of semitechnical to technical papers presented at a conference in 1970. Topics covered include prevention, effects of oil and chemicals used to cleanup oil on plant and animal life and on treatment plants, analysis, and case studies.

Hoult, David P., ed. OIL ON THE SEA. New York: Plenum Press, 1969. vii, 114 p.

Proceedings from a symposium which covered, in general, the scientific and engineering aspects of oil pollution. Best articles deal with methods of oil spill cleanup.

INTERNATIONAL CONFERENCE ON OIL POLLUTION OF THE SEA. Winchester, Engl.: Warren and Sons, Wykeham Press, 1969. 414 p.

Collection of papers ranging in scope from legal and administrative aspects of oil pollution to biological effects of oil on marine flora and fauna.

Kolpack, Ronald L., ed. BIOLOGICAL AND OCEANOGRAPHIC SURVEY OF THE SANTA BARBARA CHANNEL OIL SPILL (1969-70). Vol. 2.: PHYSICAL, CHEMICAL AND GEOLOGICAL STUDIES. Los Angeles: University of Southern California, Alan Hancock Foundation, 1971. 477 p.

Contains data on the physical characteristics of the area involved in the Santa Barbara oil spill, includes transport and deposition of sediments, as well as hydrocarbon content of sediments. Volume 1 covers the effects of the oil spill on the biota (see Straughan, p. 98).

Marx, Wesley. OILSPILL. Sierra Club Battlebook Series. San Francisco: Sierra Club, 1971. 139 p.

Popular treatment of the problem of oil and oil spill accounts the potential havoc, the effects on marine life, people, and the oil industry.

Nelson-Smith, Anthony. "Effects of the Oil Industry on Shore Life in Estuaries." PROCEEDINGS OF THE ROYAL SOCIETY OF LONDON,Series B, 180 (March 1972): 487-96.

Overview of oil pollution includes in the discussion how oil enters the water system, the toxicity of the various components of oil, and some solutions to the problem.

______. OIL POLLUTION AND MARINE ECOLOGY. New York: Plenum Publishing, 1972. ix, 260 p.

Survey of oil pollution and its effects on marine ecology. Discusses the sources and properties of spilled oil including toxicity, effects of oil on marine organisms and communities, as well as the economic factors.

OIL SPILLAGE STUDY: LITERATURE SEARCH AND CRITERIA TO CONTROL AND PREVENT DAMAGE. Washington, D.C.: Defense Documentation Center, Defense Supply Agency, 1967. Distributed by National Technical Information Service, U.S. Department of Commerce, Springfield, Va. Various pagination.

Deals with the technical aspects of oil spill prevention and control. Contains many references.

Potter, Jeffrey. DISASTER BY OIL; OIL SPILLS: WHY THEY HAPPEN, WHAT THEY DO, HOW WE CAN END THEM. New York: Macmillan, 1973. xii, 301 p.

Detailed account of four oil spills from the Torrey Canyon to the Santa Barbara, stresses what happened, how it was handled and the actions taken (judicial). Does not go into biological effects.

Schwartzberg, Henry G. SPREADING AND MOVEMENT OF OIL SPILLS. Washington, D.C.: Federal Water Pollution Control Administration, Department of the Interior, 1970. vi, 103 p.

Technical review of the physical parameters involved in the movement of oil on water.

Smith, James E. "TORREY CANYON" POLLUTION AND MARINE LIFE. Cambridge, Engl.: Cambridge University Press, 1968. xiv, 196 p.

Examines the toxic effects of the oil spilled from the Torrey Canyon and of the detergents used in cleanup on aquatic life (generally intertidal plants and animals). Presents survey data as well as results from toxicity experiments.

Steinhart, Carol E., and Steinhart, John S. BLOWOUT: A CASE STUDY OF THE SANTA BARBARA OIL SPILL. Belmont, Calif.: Duxbury Press, 1972. xvi, 138 p. Paper.

Concentrates primarily on what happened, i.e., the chaos of political, legal, and economic efforts to clean up the oil. Presents general summaries of the findings from the more technical reports on the effects of the oil spill.

Straughan, Dale, ed. BIOLOGICAL AND OCEANOGRAPHIC SURVEY OF THE SANTA BARBARA CHANNEL OIL SPILL (1969-70). Vol. 1.: BIOLOGY AND BACTERIOLOGY. Los Angeles: University of Southern California, Alan Hancock Foundation, 1971. 426 p.

Series of articles which indicate the effects of the oil spill on the biota (bacteria to whales) of Santa Barbara. Volume 2 covers the physical, chemical, and geological aspects (see Kolpack, p. 97).

U.S. Environmental Protection Agency. WATER POLLUTION POTENTIAL OF SPENT OIL SHALE RESIDUES. Water Pollution Control Research Series 14030 EDB. Washington, D.C.: 1971. 116 p.

Report of a study which examined the concentration of spent oil shale in runoff and correlated this concentration with runoff rate, precipitation intensity, and flow, to determine residence time in streams. Technical.

3. Journal Articles

Atlas, R.M.; Schofield, E.A.; Morelli, E.A.; and Cameron, P.E. "Effects of Petroleum Pollutants on Arctic Microbial Populations." ENVIRONMENTAL POLLUTION 10 (January 1976): 35-43.

Ayers, R.C., Jr.; Jahns, H.E.; and Glaeser, J.L. "Oil Spills in the Arctic Ocean: Extent of Spreading and Possibility of Large Scale Thermal Effects." SCIENCE 186 (20 November 1974): 843-44.

Blumer, M[ax]. "Scientific Aspects of the Oil Spill Problem." ENVIRONMENTAL AFFAIRS 1 (April 1971): 54-73.

Clark, R.C., Jr., and Finley, J.S. "Tidal Aquarium for Laboratory Studies of Environmental Effects on Marine Organisms." PROGRESSIVE FISH CULTURALIST 36 (July 1974): 134-37.

Clark, R.C., Jr.; Finley, J.S.; and Gibson, G.G. "Acute Effects of Outboard Motor Effluent on Two Marine Shellfish." ENVIRONMENTAL SCIENCE AND TECHNOLOGY 8 (November 1974): 1009-14.

Dennington, V.N.; George, J.J.; and Wyborn, C.H.E. "The Effects of Oils on Growth of Freshwater Phytoplankton." ENVIRONMENTAL POLLUTION 8 (April 1975): 233-37.

Evans, D.R., and Rice, S.D. "Effects of Oil on Marine Ecosystems: A Review for Administrators and Policymakers." NATIONAL MARINE FISHERIES SERVICE FISHERY BULLETIN 72 (July 1974): 625-38.

Golden, P.C. "Oil Removal Techniques in an Arctic Environment." JOURNAL MARINE TECHNOLOGY SOCIETY 8 (January 1974): 38-43.

Gordon, D.C., Jr., and Prouse, N.J. "The Effects of Three Oils on Marine Phytoplankton Photosynthesis." MARINE BIOLOGY 22 (November 1973): 329-33.

Hay, K.G. "Oil and the Sea--The Ecological Implications of a Controversial Invasion." JOURNAL MARINE TECHNOLOGY SOCIETY 8 (January 1974): 19-20.

James, W.P.; Hann, R.W., Jr.; and Slowey, J. "Effects of an Offshore Crude Oil Unloading Terminal on the Marine Environment." JOURNAL MARINE TECHNOLOGY SOCIETY 9 (January 1975): 27-31.

Klein, A.E. "The Effects of Artificial Sunlight Upon Floating Oils." WATER RESEARCH 8 (January 1974): 79-83.

Lee, R.F.; Sauerheker, R.; and Benson, A.A. "Petroleum Hydrocarbons: Uptake and Discharge by the Marine Mussel Mytilus Edulis." SCIENCE 177 (28 July 1972): 344-46.

Linden, O. "Effects of Oil on the Amphipod Gammarus Oceanicus." ENVIRONMENTAL POLLUTION 10 (June 1976): 239-50.

Livingston, D. "Oil on the Seas." ENVIRONMENT 16 (September 1974): 38-43.

Morris, R.J. "Uptake and Discharge of Petroleum Hydrocarbons by Barnacles." MARINE POLLUTION BULLETIN 4 (July 1973): 107-9.

Neff, J.M., and Anderson, J.W. "An Ultraviolet Spectophotometric Method for the Determination of Naphthalene and Alkylnaphalenes in the Tissues of Oil-Contaminated Marine Animals. BULLETIN OF ENVIRONMENTAL CONTAMINATION AND TOXICOLOGY 14 (July 1975): 122-28.

Prouse, N.J.; Gordon, D.C., Jr.; and Keizer, P.D. "Effects of Low Concentrations of Oil Accommodated in Sea Water on the Growth of Unialgal Marine Phytoplankton Cultures." JOURNAL FISHERIES RESEARCH BOARD OF CANADA 33 (April 1976): 810-18.

Swinnerton, J.W., and Lamontagne, R.A. "Oceanic Distribution of Low-Molecular Weight Hydrocarbons: Baseline Measurements." ENVIRONMENTAL SCIENCE AND TECHNOLOGY 8 (July 1974): 657-63.

Tagatz, M.E. "Reduced Oxygen Tolerance and Toxicity of Petroleum Products to Juvenile American Shad." CHESAPEAKE SCIENCE 2 (March-June 1961): 65-70.

Walker, J.D.; Colwell, R.R.; Vaituzis, Z.; and Meyer, S.A. "Petroleum Degrading Achlorophyllous Alga Prototheca Zopfii." NATURE 254 (3 April 1975): 423-24.

Wells, P.G., and Sprague, J.B. "Effects of Crude Oil on American Lobster (Homarus Americanus) Larvae in the Laboratory." JOURNAL FISHERIES RESEARCH BOARD OF CANADA 33 (July 1976): 1604-14.

D. METALS AND ASSOCIATED ELEMENTS

Attention was focused on heavy metal pollution in the 1950s by the Minamata disease in Japan where people suffered mercury poisoning from eating contaminated fish. Although mercury still remains a pollution problem, primarily because of its long life within the aquatic system, it is now only one of a long list of elements which pose pollution problems. Some of the other elements are arsenic, cadium, copper, chromium, lead, magnesium, and zinc. Early research on heavy metal pollution was concerned primarily with measuring the element in the environment, and with its toxicity. Research today involves determining and mathematically describing the reactions of these elements with other dissolved components in the water, with the sediment, and with the aquatic biota. One of the new areas of research is the use of model ecosystems to describe the movement and transformation of the element in the environment.

Mercury has predominated the literature of the 1960s and 1970s, with other elements enjoying transient popularity as their presence becomes detectable. The most recent information on heavy metal pollution will be found in journal articles; those listed in this section will introduce the reader to this literature. Additional references may be found under the section on industrial pollution, p. 133.

1. Abstracts and Bibliographies

Caton, G.M. MERCURY IN THE ENVIRONMENT: AN ANNOTATED BIBLIOGRAPHY. Oak Ridge, Tenn.: Oak Ridge National Laboratory, 1972. vi, 271 p.

> Contains more than 550 citations alloted to various aspects of mercury contamination ranging from its chemistry to mining and pollution. Toxicological and legal aspects are also included.

U.S. Department of the Interior. Office of Library Services. MERCURY CONTAMINATION IN THE NATURAL ENVIRONMENT. Washington, D.C.: 1970. vi, 32 p.

> Bibliographic citations of English-language material relating to mercury and fish, wildlife, and water pollution (1960-70).

[U.S. Department of Interior.] Water Resources Scientific Information Center. ARSENIC AND LEAD IN WATER: A BIBLIOGRAPHY. Bibliography Series WRSIC 71-209. Washington, D.C.: 1971. Distributed by National Technical Information Service. Unpaged.

> Citations from SELECTED WATER RESOURCES ABSTRACTS through 1970 on the presence and effect of arsenic and lead in water.

______. CADMIUM IN WATER: A BIBLIOGRAPHY. Bibliography Series WRSIC 73-209. Washington, D.C.: 1973. Distributed by National Technical Information Service. 231 p.

Citations from SELECTED WATER RESOURCES ABSTRACTS through 15 December 1972 on all aspects of cadmium in water.

______. CHROMIUM IN WATER: A BIBLIOGRAPHY. Bibliography Series WRSIC 72-205. Washington, D.C.: 1972. Distributed by National Technical Information Service. iv, 126 p.

Annotated bibliography pertaining to chromium in water from SELECTED WATER RESOURCES ABSTRACTS through 15 May 1972.

______. COPPER IN WATER: A BIBLIOGRAPHY. Bibliography Series WRSIC 71-204. Washington, D.C.: 1971. Distributed by National Technical Information Service. 189 p.

Citations from SELECTED WATER RESOURCES ABSTRACTS through 15 June 1971 on all aspects of the relationship between copper and water.

______. MAGNESIUM IN WATER: A BIBLIOGRAPHY. Bibliography Series WRSIC 71-206. Washington, D.C.: 1971. Distributed by National Technical Information Service. 152 p.

Citations from SELECTED WATER RESOURCES ABSTRACTS through 15 June 1971 on the presence of magnesium in water and its effects on the biota.

______. MANGANESE IN WATER: A BIBLIOGRAPHY. Bibliography Series WRSIC 71-205. Washington, D.C.: 1971. Distributed by National Technical Information Service. 127 p.

Citations from SELECTED WATER RESOURCES ABSTRACTS through 15 June 1971 on the occurrence of manganese in water.

______. MERCURY IN WATER: A BIBLIOGRAPHY. 2 vols. Bibliography Series WRSIC 72-214. Washington, D.C.: 1972, 1975. Distributed by National Technical Information Service.

Citations from SELECTED WATER RESOURCES ABSTRACTS from 1967 through 15 December 1971 for volume 1 and from 1972 through 15 February 1975 for volume 2. All aspects of mercury are covered from analysis to toxicity.

______. TRACE ELEMENTS IN WATER: A BIBLIOGRAPHY. Bibliography Series WRSIC 71-202. Washington, D.C.: 1971. Distributed by National Technical Information Service. 287 p.

Citations from SELECTED WATER RESOURCES ABSTRACTS through 15 June 1971, Trace elements covered include arsenic, barium, beryllium, boron, molybdenum, antimony, bismuth, cobalt, nickel, selenium, silicon, silver, and titanium.

______. ZINC IN WATER: A BIBLIOGRAPHY. Bibliography Series WRSIC 71-208. Washington, D.C.: 1971. Distributed by National Technical Information Service. 138 p.

Citations from SELECTED WATER RESOURCES ABSTRACTS through 15 June 1971, covering all aspects of zinc in water.

2. References

Afghan, B.K. ANALYSIS OF WATER CONTAMINANTS WITH SPECIAL REFERENCE TO HEAVY METALS. Technical Bulletin, no. 52. Ottawa: Department of the Environment, Inland Waters Branch, 1971. 16 p. Pamphlet.

General summary of methods useful in measuring heavy metals, i.e., what the methods entail, for which metals they work best, and the sensitivities and detection limits.

Button, D.K., and Dunker, S.S. BIOLOGICAL EFFECTS OF COPPER AND ARSENIC POLLUTION. Institute of Marine Science, University of Alaska Report, no. R71-8. College: University of Alaska, 1971. 59 p.

Final report of a study on the effects of copper and arsenic on marine microorganisms. Found that the toxicity levels of antimetabolites depend not only on the organism but in a major way on free nutrients and trace metal levels.

Davies, Patrick H. EFFECTS OF CHEMICAL VARIATIONS IN AQUATIC ENVIRONMENTS. Vol. 3: LEAD TOXICITY TO RAINBOW TROUT AND TESTING APPLICATION FACTOR CONCEPT. Ecological Research Series, R3-73-011C. Washington, D.C.: U.S. Environmental Protection Agency, 1973. viii, 80 p.

Report of a study on acute and chronic lead toxicities to rainbow trout and on the validity of the application factor concept using two types of water (hard well water and soft dechorinated tap water).

D'itri, Frank M. THE ENVIRONMENTAL MERCURY PROBLEM: A REPORT TO THE MICHIGAN HOUSE OF REPRESENTATIVES RESULTING FROM HOUSE RESOLUTION 424, GREAT LAKES CONTAMINATION. Michigan State University, Institute of Water Research Technical Report, no. 12. Lansing: House of Representatives, 1971. xiii, 289 p.

Comprehensive report on mercury pollution including data on background concentrations, methods of determination, and suggestions for removal and/or decontamination of polluted areas. Effect of mercury on food chains is also examined.

Everhart, W.H. EFFECTS OF CHEMICAL VARIATION IN AQUATIC ENVIRONMENTS. Vol. 2: TOXIC EFFECTS OF AQUEOUS ALUMINUM TO RAINBOW TROUT. Ecological Research Series, R3-73-011B. Washington, D.C.: U.S. Environmental Protection Agency, 1973. vii, 41 p.

> Report of an investigation on the effects of soluble anionic species and neutral precipitates of aluminum hydroxide and lead nitrate complexes on fingerling trout and on the fertilization of eggs, and early life history.

Feich, George. CONTROL OF MERCURY CONTAMINATION IN FRESHWATER SEDIMENTS. Environmental Protection Technical Series R2-72-077. Washington, D.C.: Environmental Protection Agency, 1972. Distributed by U.S. Government Printing Office. 156 p.

> Report of a study which developed and evaluated physical and chemical methods for binding mercury to the sediments and thereby preventing its release.

Flick, D.F., et al. "Toxic Effects of Cadmium: A Review." ENVIRONMENTAL RESEARCH 4 (April 1971): 71-85.

> Review includes epidemiology, distribution, industrial production, and uses of cadmium, as well as modes of incorporation into the human body and possible significance with respect to metabolism.

Friberg, Lars; Piscator, Magnus; and Nordberg, G. CADMIUM IN THE ENVIRONMENT. Cleveland: CRC Press, 1971. 166 p.

> Focuses on information needed to understand the toxic action of cadmium, and the relationship between dose and effects on human beings and animals.

Fromm, Paul O. TOXIC ACTION OF WATER SOLUBLE POLLUTANTS ON FRESHWATER FISH. Washington, D.C.: U.S. Environmental Protection Agency, Water Quality Office, 1970. viii, 56 p.

> Report on the effects of various pollutants (chromium, ammonia) on trout and goldfish. Responses monitored included plasma cortisol levels and liver glycogen stores. Also cites studies on the uptake of various compounds, including pesticides by the gills.

Gavis, Jerome, and Ferguson, John F. "The Cycling of Mercury Through the Environment." WATER RESEARCH 6 (September 1972): 989-1008.

> Review of the mercury cycle in the aquatic environment with emphasis on the aquatic chemistry of mercury.

Hartung, Rolf, and Dinman, Bertram, eds. ENVIRONMENTAL MERCURY CONTAMINATION. Ann Arbor, Mich.: Ann Arbor Science Publishers, 1972. ix, 349 p.

> Based on a conference held in 1970 which examined the occurrence of mercury, its analysis, environmental dynamics, and biological effects. Good review up to 1970.

Jernelov, Arne, and Lann, H. "Mercury Accumulation in Food Chains." OIKOS 22, no. 3 (1971): 403-6.

> Mercury accumulation in the food chain was found to be a function of levels in both prey organisms and in the water.

Katz, Arthur. "Mercury Pollution: The Making of an Environmental Crisis." CRC CRITICAL REVIEWS IN ENVIRONMENTAL CONTROL 2 (January 1972): 517-34.

> Review of the socioadministrative aspects of mercury pollution, as well as the processes of accumulation and physiological effects on aquatic organisms.

Kopp, John F., and Kroner, Robert C. TRACE METALS IN WATER OF THE UNITED STATES: A FIVE-YEAR SUMMARY OF TRACE METALS IN RIVERS AND LAKES OF THE UNITED STATES. Cincinnati: U.S. Department of Interior, Federal Water Pollution Control Administration, Division of Pollution Surveillance, 1967. 200 p.

> Provides data for various river systems between October 1962 and October 1967 including detection and measurement of eighteen elements from arsenic to zinc.

Mathis, B.J., and Kevern, N.R. DISTRIBUTION OF MERCURY, CADMIUM, LEAD AND THALLIUM IN A EUTROPHIC LAKE. Institute of Water Research Technical Report, no. 34. East Lansing: Michigan State University, 1973. v, 25 p.

> Report of a study which examined the concentrations of the title compounds in the sediment and water of a lake, as well as five species of fish, two species of aquatic macrophytes, zooplankton, and the feces of migratory geese inhabiting the lake.

New York State University. CHROMIUM, CADMIUM, ARSENIC, SELENIUM, MERCURY AND AQUATIC LIFE: A BRIEF LITERATURE REVIEW. Great Lakes Laboratory Special Report, no. 9. Buffalo: 1971. ii, 23 p.

> Survey of the literature relating the effects of heavy metals on aquatic life up to 1970.

Nilsson, R. ASPECTS ON THE TOXICITY OF CADMIUM AND ITS COMPOUNDS: A REVIEW. Stockholm: Swedish Natural Science Research Council, 1969. 59 p.

> Literature review on the occurrence and the biochemical and toxicological effects of cadmium and its compounds. The occurrence of cadmium in water, foods, vegetation, fertilizer, and humans is discussed along with its chemical properties, uses and toxic effects.

Saha, J.G. "Significance of Mercury in the Environment." RESIDUE REVIEWS 42 (1972): 103-63.

> Literature review dealing with the contamination of the environment by residues of mercury. Included are the occurrence of mercury in water and the chemical nature and toxicity of methlmercury compounds.

Singer, Philip C., ed. TRACE METALS AND METAL-ORGANIC INTERACTIONS IN NATURAL WATERS. Ann Arbor, Mich.: Ann Arbor Science Publishers, 1973. x, 380 p.

> Collection of papers presented at a symposium dealing primarily with the chemistry of trace metals including analysis and factors affecting distribution. Technical.

Suggs, James D.; Petersen, Donald H.; and Middlebrook, James B., Jr. MERCURY POLLUTION CONTROL IN STREAM AND LAKE SEDIMENTS. Water Pollution Control Research Series 16080 HTD. Washington, D.C.: U.S. Environmental Protection Agency, 1972. vi, 38 p.

> Results of a study on mercury removal from sediments without destroying the environment based on two thousand mercury analyses covering a wide range of "getter" combinations and environmental conditions.

Whitton, B.A. "Toxicity of Heavy Metals to Freshwater Algae: A Review." PHYKOS 9, no. 2 (1970): 116-25.

> Review of the literature on the toxicity of heavy metals; full spectrum to freshwater algae. Account is mainly interested in field studies.

Wood, J.M. "Environmental Pollution by Mercury." In ADVANCES IN ENVIRONMENTAL SCIENCE AND TECHNOLOGY, edited by James N. Pitts, Jr. and Robert L. Metcalf, vol. 2, pp. 39-56. New York: Wiley-Interscience, 1971.

> Review of mercury pollution up to 1971, especially the conversion of inorganic to organic. Covers sources, history, magnitude of problem, and possible solutions.

3. Journal Articles

Bakir, F., et al. "Methylmercury Poisoning in Iraq." SCIENCE 181 (July 1973): 230-41.

Barber, R.T., and Vijayakumar, A. "Mercury Concentrations in Recent and Ninety-Year-Old Benthopelagic Fish." SCIENCE 178 (10 November 1972): 636-38.

Brown, B., and Ahsanullah, M. "Effect of Heavy Metals on Mortality and Growth." MARINE POLLUTION BULLETIN 2 (December 1971): 182-87.

Bruland, K.W.; Bertine, K.; Koide, M.; and Goldberg, E.D. "History of Metal Pollution in Southern California Coastal Zone." ENVIRONMENTAL SCIENCE AND TECHNOLOGY 8 (May 1974): 425-32.

Burrows, W.D., and Krenkel, P.A. "Studies on Uptake and Loss of Methylmercury-203 by Bluegills (Lepomis Macrochirus Raf)." ENVIRONMENTAL SCIENCE AND TECHNOLOGY 7 (December 1973): 1127-30.

Chau, Y.K.; Wong, P.T.S.; Silverberg, B.A.; Luxon, P.L.; and Bengert, G.A. "Methylation of Selenium in the Aquatic Environment." SCIENCE 192 (11 June 1976): 1130-31.

Goldwater, L.J. "Mercury in the Environment." SCIENTIFIC AMERICAN 224 (May 1971): 15-21.

Hall, S.K. "Lead Pollution and Poisoning." ENVIRONMENTAL SCIENCE AND TECHNOLOGY 6 (January 1972): 30-35.

Jernelov, A., and Lann, H. "Studies in Sweden on Feasibility of Some Methods for Restoration of Mercury-Contaminated Bodies of Water." ENVIRONMENTAL SCIENCE AND TECHNOLOGY 7 (August 1973): 712-18.

Kinnison, R.R. "Pb: In Search of the Facts." ENVIRONMENTAL SCIENCE AND TECHNOLOGY 10 (July 1976): 644-49.

Guide to a compilation of information dealing with the biological impact of lead on man.

Klein, D.H., and Goldberg, E.D. "Mercury in the Marine Environment." ENVIRONMENTAL SCIENCE AND TECHNOLOGY 4 (September 1970): 765-68.

Kubota, J.; Mills, E.L.; and Oglesby, R.T. "Lead, Cd, Zn, Cu, and Co in Streams and Lake Waters of Cayuga Lake Basin, New York." ENVIRONMENTAL SCIENCE AND TECHNOLOGY 8 (March 1974): 243-48.

Lindberg, S.E., and Harriss, R.C. "Mercury-Organic Matter Associations in Estuarine Sediments and Interstitial Water." ENVIRONMENTAL SCIENCE AND TECHNOLOGY 8 (May 1974): 459-62.

Miller, G.E., et al. "Mercury Concentrations in Museum Specimens of Tuna and Swordfish." SCIENCE 175 (10 March 1972): 1121-22.

Morel, F.M.M.; Westall, J.C.; and Omelia, C.R. "Fate of Trace Metals in Los Angeles County Wastewater Discharge." ENVIRONMENTAL SCIENCE AND TECHNOLOGY 9 (August 1975): 756-61.

Patrick, F.M., and Loutit, M. "Passage of Metals in Effluents Through Bacteria to Higher Organisms." WATER RESOURCES 10 (April 1976): 333-35.

Potter, L.; Kidd, D.; and Standiford, D. "Mercury Levels in Lake Powell: Bioamplication in Man-Made Desert Reservoir." ENVIRONMENTAL SCIENCE AND TECHNOLOGY 9 (January 1975): 41-46.

Putman, J.J., and Madden, R.W. "Quicksilver and Slow Death." NATIONAL GEOGRAPHIC SOCIETY 142 (October 1972): 506-27.

Selikoff, I.S., ed. "Hazards of Mercury." ENVIRONMENTAL RESEARCH 4 (March 1971): 1-69.

Siegel, S.M., and Eshleman, A. "Chlorine Bleaches: A Significant Long Term Source of Mercury Pollution." WATER, AIR AND SOIL POLLUTION 4 (July-August 1975): 461-63.

Weiss, H.V.; Koide, M.; and Goldberg, E.D. "Mercury in the Greenland Ice Sheet: Evidence of Recent Input by Man." SCIENCE 174 (12 November 1971): 692-94.

Wood, J.M. "Biological Cycles for Toxic Elements in the Environment." SCIENCE 183 (March 1974): 1049-52.

E. THERMAL POLLUTION

The primary source of thermal pollution is power plants. Other sources are industrial waste discharges and the removal of plant cover from streams during logging operations. Research on thermal pollution has focused both on the engineering aspects associated with power plant operation and on the effects of heated effluents on the aquatic biota. Some effects associated with higher ambient temperatures include increased mortality and/or emigration of species, both of which result in changes in community composition. Elevated temperatures also influence an organism's response to environmental factors as salinity, oxygen concentration, and toxicants. There are, however, beneficial uses for waste heat from power plants, especially in crop production and aquaculture, the growing of plants and animals in water. Heated water can also be used by industry and for building heating.

The listed journal articles are indicative of both the scope of current research in this area and of the journals containing thermal pollution literature. Additional references on thermal pollution are in the section on industrial pollution, p. 133.

1. Abstracts and Bibliographies

Raney, Edward C., and Menzel, Bruce W. HEATED EFFLUENTS AND EFFECTS ON AQUATIC LIFE WITH EMPHASIS ON FISH. 3d ed. Washington, D.C.: Office of Water Resources Research, Department of the Interior, 1969. 470 p.

> Bibliographic citations of the literature concerned with the effects of heat on aquatic organisms, primarily fish. More than eighteen hundred references are given.

2. References

Brown, George W., and Brazier, Jon R. CONTROLLING THERMAL POLLUTION IN SMALL STREAMS. U.S. Environmental Protection Agency Technical Series R2-72-083. Washington, D.C.: U.S. Environmental Protection Agency, 1972. vii, 64 p.

> Report on the effect of leaving buffer strips along streams during logging operations on stream temperature.

Cairns, John, Jr., and Lanza, Guy R. THE EFFECTS OF HEATED WASTE WATERS ON SOME MICROORGANISMS. Blacksburg: Virginia Polytechnic Institute and State University, Water Resources Center, 1972. x, 101 p.

> Report of a study on the effects of heated waters on protozoans, algae and some fungi. Two categories of thermal shock were examined: simulation of passage through a condensing system, and effects of entrainment upon the microbes below the discharge points.

Christianson, Alden G., and Tichenor, Bruce A. INDUSTRIAL WASTE GUIDE ON THERMAL POLLUTION. Corvallis, Oreg.: Pacific Northwest Water Laboratory, 1968. vii, 112 p.

> Reviews the output of heated water by various industries (i.e., electric power, manufacturing) and the effects of heated discharges on the environment, physical, chemical and biological. Technical.

Cole, Lamont C. "Thermal Pollution." In MAN'S IMPACT ON ENVIRONMENT, edited by Thomas R. Detwyler, pp. 217-24. New York: McGraw-Hill, 1971.

Briefly discusses sources of waste heat, biological effects, and predictions for the future.

Copeland, B.J., and Davis, H.L. ESTUARINE ECOSYSTEMS AND HIGH TEMPERATURE. North Carolina Water Resources Research Institute Report, no. 68. Raleigh: North Carolina State University, 1972. ix, 90 p.

Report of a study which investigated the response of estuarine community structure, respiration, and production to added heat, sewage, and their combinations. Organisms included in the study were phytoplankton, grass crabs, shrimp, clams, and fish.

Gammon, James R. THE EFFECT OF THERMAL INPUTS ON THE POPULATIONS OF FISH AND MACROINVERTEBRATES IN THE WABASH RIVER. Water Resources Research Center Technical Report, no. 32. West Lafayette, Ind.: Purdue University, 1973. Paper. x, 106 p.

Report of study which covered six years on the Wabash River near two electrical generating plants. Fish and macroinvertebrate populations were examined. Study found behavioral changes in the fish species which allowed them to select water of an optimum temperature.

Jensen, Loren D.; Davies, Robert M.; Brooks, Arthur S.; and Meyers, Caldwell, D. THE EFFECTS OF ELEVATED TEMPERATURE UPON AQUATIC INVERTEBRATES: A REVIEW OF THE LITERATURE RELATING TO FRESHWATER AND MARINE INVERTEBRATES. Edison Electric Institute Research Project Report, no. 49. New York: Edison Electric Institute, 1969. xii, 232 p.

Literature review covering both the general role of temperature in aquatic ecosystems as well as the effect on each individual invertebrate group such as insects, crustacea, and mollusks. Includes a discussion of the effects of temperature on the physical-chemical portion of the aquatic ecosystem.

Krenkel, Peter A., and Parker, Frank L., eds. BIOLOGICAL ASPECTS OF THERMAL POLLUTION. Nashville, Tenn.: Vanderbilt University Press, 1969. xx, 407 p.

Papers from a symposium on thermal pollution which covered primarily the effects of thermal pollution on such aquatic organisms as marine zooplankton, marine and estuarine benthos, fish, algae, and freshwater benthos.

Mihursky, Joseph A. PATUZENT THERMAL STUDIES. Natural Resources Institute Special Report, no. 1. Baltimore: University of Maryland, 1969. 20 p.

Summary of the results of a study on the temperature changes induced by the Patuxent power plant in which references are given for more complete results. Contents include field and laboratory study results, predictions from laboratory data, and evaluation of Maryland's temperature regulations.

Mihursky, J[oseph]. A., et al. EFFECTS OF THERMAL POLLUTION ON PRODUCTIVITY AND STABILITY OF ESTUARINE COMMUNITIES. College Park: Maryland University, Water Resources Research Center, 1971. 65 p.

Field and laboratory studies were utilized to provide data concerning the environmental effects of an electrical generating facility. Suggestions are included for alternative operating procedures and engineering designs.

National Wildlife Federation. THERMAL POLLUTION AND COOLING ALTERNATIVES; SUPPLEMENT TO NUCLEAR POWER PLANTS AND YOU--A CITIZENS HANDBOOK. Washington, D.C.: National Wildlife Federation, 1971. 30 p.

General explanation of the thermal pollution problem and what is, or can be done to solve the problem.

Parker, Frank L., and Krenkel, Peter A. PHYSICAL AND ENGINEERING ASPECTS OF THERMAL POLLUTION. Cleveland: CRC Press, 1970. 100 p.

Discussion of the mechanics of heated wastewater discharge. Contents include effects of heated discharges on waste assimilation, beneficial effects of heat additions, mechanisms of discharges, modeling, and methods of cooling with a comparison of methods.

______, eds. ENGINEERING ASPECTS OF THERMAL POLLUTION. Nashville: Vanderbilt University Press, 1969. xxiv, 351 p.

Proceedings of a national symposium on thermal pollution held in Nashville on engineering aspects of thermal pollution. Covered such areas as modeling, power plant construction, as well as physical effects on the environment and economic considerations.

U.S. Atomic Energy Commission. Division of Reactor Development and Technology. THERMAL EFFECTS AND U.S. NUCLEAR POWER STATIONS. Washington, D.C.: 1971. v, 40 p.

Presents the problems facing utilities, and federal and state agencies concerned with the thermal effects associated with the production of electrical power. Includes a general overview of the methods of heat disposal, beneficial uses of waste heat, ecological aspects of thermal discharges, and water temperature standards.

U.S. Office of Water Quality. RESEARCH ON THE PHYSICAL ASPECTS OF THERMAL POLLUTION. Water Pollution Control Research Series 16130 DPUO2 171. Washington, D.C.: U.S. Environmental Protection Agency, 1971. Distributed by U.S. Government Printing Office. 189 p.

> Report of the results of a study on the mechanisms of formation and maintenance of the thermal structure in deep temperature lakes, and the effects on this structure of discharges of waste heat from electrical generating plants. Technical.

3. Journal Articles

Axtmann, R.C. "Environmental Impact of a Geothermal Power Plant." SCIENCE 187 (7 March 1975): 795-803.

Becker, C.D. "Columbia River Thermal Effects Study: Reactor Effluent Problems." JOURNAL WATER POLLUTION CONTROL FEDERATION 45 (May 1973): 850-68.

Brown, George W., and Krygier, J.T. "Effects of Clear-Cutting on Stream Temperature." WATER RESOURCES RESEARCH 6 (August 1970): 1133-39.

Bush, R.M.; Welch, E.B.; and Mar, B.W. "Potential Effects of Thermal Discharges on Aquatic Systems." ENVIRONMENTAL SCIENCE AND TECHNOLOGY 8 (June 1974): 561-68.

Cairns, J[ohn]., Jr. "Coping with Heated Waste Discharges from Steam-Electric Power Plants." BIOSCIENCE 22 (July 1972): 411-19.

______. "Ecological Management Problems Caused by Heated Waste Water Discharge into the Aquatic Environment." WATER RESOURCES BULLETIN 6 (November-December 1970): 868-78.

______. "Thermal Pollution--A Cause for Concern." JOURNAL WATER POLLUTION CONTROL FEDERATION 43 (January 1971): 55-66.

Cairns, J[ohn]., Jr.; Heath, A.G.; and Parker, B.C. "Temperature Influence on Chemical Toxicity to Aquatic Organisms." JOURNAL WATER POLLUTION CONTROL FEDERATION 47 (February 1975): 267-80.

Cherry, D.S.; Guthrie, R.K.; and Harvey, R.S. "Temperature Influence on Bacterial Populations in Three Aquatic Systems." WATER RESEARCH 8 (March 1974): 149-55.

Clark, J.R. "Thermal Pollution and Aquatic Life." SCIENCE AMERICAN 220 (March 1969): 18-27.

Coutant, C.C. "Biological Aspects of Thermal Pollution. II--Scientific Basis for Water Temperature Standards at Power Plants." CRC CRITICAL REVIEWS IN ENVIRONMENTAL CONTROL 3 (August 1972): 1-24.

______. "How to Put Waste Heat to Work." ENVIRONMENTAL SCIENCE AND TECHNOLOGY 10 (September 1976): 868-71.

Gray, J.R.A., and Edington, J.M. "Effects of Woodland Clecrance on Stream Temperature." JOURNAL FISHERIES RESEARCH BOARD OF CANADA 26 (February 1969): 399-403.

Hansmann, Eugene W., and Phinney, H.K. "Effects of Logging on Periphyton in Coastal Streams of Oregon." ECOLOGY 54 (Winter 1973): 194-99.

Lee, S.S.; Vezirogiu, T.N.; Sengupta, S.S.; and Weinberg, N.L. "Remote Sensing Applied to Thermal Pollution." In ADVANCES IN THERMAL ENGINEERING, edited by J.P. Hartnett and T.F. Irvine, Jr., vol. 5, pp. 303-34. Washington, D.C.: Hemisphere Publishing, 1975.

Levin, A.A.; Birch, T.J.; Hillman, R.E.; and Rains, G.E. "Thermal Discharges: Ecological Effects." ENVIRONMENTAL SCIENCE AND TECHNOLOGY 6 (March 1972): 224-30.

Logan, D.T., and Maurer, D. "Diversity of Marine Invertebrates in a Thermal Effluent." JOURNAL WATER POLLUTION CONTROL FEDERATION 47 (March 1975): 515-23.

Merriman, D. "Calefaction of a River." SCIENTIFIC AMERICAN 222 (May 1970): 45-52.

Moore, F.K., and Jaluria, Y. "Thermal Effects of Power Plants on Lakes." Transactions of American Society of Mechanical Engineering, Series C. JOURNAL OF HEAT TRANSFER 94 (May 1972): 163-68.

O'Sullivan, A.J., and Collinson, R.I. "Wasted Heat and Aquatic Ecosystems." EFFLUENT AND WATER TREATMENT JOURNAL 16 (January 1976): 15-17.

Parker, E.D.; Hershfield, M.F.; and Gibbons, J.W. "Ecological Comparisons of Thermally Affected Aquatic Environments." JOURNAL WATER POLLUTION CONTROL FEDERATION 45 (April 1973): 726-33.

Selbergeld, E.K. "Dieldrin Effects of Chronic Sublethal Exposure on Adaptation to Thermal Stress in Freshwater Fish." ENVIRONMENTAL SCIENCE AND TECHNOLOGY 7 (September 1973): 846-49.

Tarzwell, C.M. "An Argument for the Open Ocean Siting of Coastal Thermal Electric Plants." JOURNAL ENVIRONMENTAL QUALITY 1 (January-March 1972): 89-91.

F. SUSPENDED MATTER

Although it has long been recognized that excessive suspended matter in natural bodies of water can be detrimental to aquatic organisms, it was not acknowledged as a major pollutant until recently. Suspended matter can, for example, obstruct light interfering with or inhibiting photosynthetic activity, interfere with feeding processes, damage tissues, asphyxiate some organisms, and damage certain life stages. Suspended matter is also important because it may react chemically with other pollutants such as heavy metals and pesticides, affecting their form, distribution, and toxicity.

Many of the following references are concerned with the effect of suspended matter on aquatic life. The reader should check the sections on metals, p. 101, and pesticides, p. 83, for additional references on the role of suspended matter in the aquatic environment. Also chapter 16, "Pollution Sources," p. 133 has references on suspended matter from agriculture, industry, and urban areas.

1. Abstracts and Bibliographies

American Society of Agricultural Engineers. EROSION AND SEDIMENT CONTROL ON URBAN AND CONSTRUCTION SITES: AN ANNOTATED BIBLIOGRAPHY. Special Publication SP 0272. St. Joseph, Mich.: 1972. 13 p. Paper.

> List of thirty-four references from the 1960s to 1972 on the above topic, including the abstract and where a copy can be obtained.

2. References

Becker, Burton C., and Mills, Thomas R. GUIDELINES FOR EROSION AND SEDIMENT CONTROL PLANNING AND IMPLEMENTATION. EPA Technology Series R2-72-015. Washington, D.C.: Environmental Protection Agency, Office of Research and Monitoring, 1972. Distributed by U.S. Government Printing Office. 228 p.

> Provides a description of how site evaluation determines the potential sediment problem and guidance for implementing a sediment and erosion control plan.

Cairns, John, Jr. "Suspended Solids Standards for the Protection of Aquatic Organisms." PURDUE UNIVERSITY. ENGINEERING EXTENSION SERIES, ENGINEERING BULLETIN no. 129, pp. 16-27. Lafayette, Ind.: Purdue University, 1968.

Review and discussion of the effects of suspended solids on aquatic organisms and proposed standards for their protection.

Carr, Donald E. DEATH OF SWEET WATERS. New York: Norton, 1966. 257 p.

Covers the silting up of rivers and lakes. Contains a large bibliography. Designed for general reader.

Cooper, A.C. THE EFFECT OF TRANSPORTED STREAM SEDIMENTS ON THE SURVIVAL OF SOCKEYE AND PINK SALMON EGGS AND ALEVIN. International Pacific Salmon Fisheries Commission Bulletin, no. 18. New Westminster, B.C.: International Pacific Salmon Fisheries Commission, 1965. 71 p.

Study of the biological effect of sediment on eggs and alevin of salmon, as well as a discussion of stream sediment transport phenomena and the physical factors affecting deposition. Technical.

Cordone, Almo, J., and Kelley, Don W. "The Influence of Inorganic Sediment on the Aquatic Life of Streams." CALIFORNIA FISH AND GAME 47 (April 1961): 189-228.

Review of the "more important" investigations made on the effects of inorganic sediment on the aquatic life of streams.

Foess, G.W., and Feng, T.H. "Bottom Deposits." JOURNAL WATER POLLUTION CONTROL FEDERATION 43 (June 1971): 1257-66.

Review of approximately eighty papers published between 1969 and 1970 on bottom deposits, topics include sampling analysis, transport phenomena, physical-chemical effects, and biological effects.

Gammon, James R. THE EFFECT OF INORGANIC SEDIMENT ON STREAM BIOTA. Water Pollution Control Research Series. Washington, D.C.: U.S. Environmental Protection Agency, Water Quality Office, 1970. Distributed by U.S. Government Printing Office. vi, 141 p.

Report of a study which evaluated the effects of sediment produced by a crushed limestone quarry on stream biota. Parameters studied were population diversity and standing crop, invertebrate drift rates, growth rates in fish, and spawning activity. Contains a good review of the literature to 1969 on the effects of sediment on aquatic biota.

Ritchie, Jerry C. "Sediment, Fish and Fish Habitat." JOURNAL OF SOIL AND WATER CONSERVATION 27 (May-June 1972): 124-25.

The effect of sediment on fish is reviewed up to 1971. Primarily a discussion-type article.

Sherk, J. Albert, Jr. THE EFFECT OF SUSPENDED AND DEPOSITED SEDIMENTS ON ESTUARINE ORGANISMS: LITERATURE SUMMARY AND RESEARCH NEEDS. Chesapeake Biological Laboratory Contribution, no. 443. Solomons: University of Maryland, Natural Resources Institute, 1971. 73 p.

Review of the literature.

3. Journal Articles

Armentano, T.V., and Woodwell, G.M. "Sedimentation Rates in a Long Island Marsh Determined by ^{210}Pb Dating." LIMNOLOGY AND OCEANOGRAPHY 20 (May 1975): 452-56.

Aston, S.R.; Thornton, I.; Webb, J.S.; Purves, J.B.; and Milford, B.L. "Stream Sediment Composition: An Aid to Water Quality Assessment." WATER, AIR AND SOIL POLLUTION 3 (September 1974): 321-25.

Benedict, P.C. "Sediment Measurement Techniques: A Fluvial Sediment." A.S.C.E. PROCEEDINGS, JOURNAL OF HYDRAULICS DIVISION 95 (September 1969): 1477-1514.

Brown, George W., and Krygier, J.T. "Clear-Cut Logging and Sediment Production in the Oregon Coast Range." WATER RESOURCES RESEARCH 7 (October 1971): 1189-98.

Burns, J.W. "Spawning Bed Sedimentation Studies in Northern California Streams." CALIFORNIA FISH AND GAME 56 (October 1970): 253-70.

Chutter, F.M. "The Effects of Silt and Sand on the Invertebrate Fauna of Streams and Rivers." HYDROBIOLOGIA 34 (1969): 57-76.

Fillos, J., and Swanson, W.R. "The Release Rate of Nutrients from River and Lake Sediments." JOURNAL WATER POLLUTION CONTROL FEDERATION 47 (May 1975): 1032-42.

Gillott, M.A.; Floyd, G.L.; and Ward, D.V. "The Role of Sediment as a Modifying Factor in Pesticide-Algae Interactions." ENVIRONMENTAL ENTOMOLOGY 4 (15 August 1975): 621-24.

Hansmann, E[ugene]. W., and Phinney, H.K. "Effects of Logging on Periphyton in Coastal Streams of Oregon." ECOLOGY 54 (Winter 1973): 194-99.

Kirchner, W.B. "An Evaluation of Sediment Trap Methodology." LIMNOLOGY AND OCEANOGRAPHY 20 (July 1975): 657-60.

Pita, F.W., and Hyne, N.J. "The Depositional Environment of Zinc, Lead and Cadmium in Reservoir Sediments." WATER RESEARCH 9 (August 1975): 701-6.

Sheridan, W.L., and McNeill W.J. "Some Effects of Logging on Two Salmon Streams in Alaska." JOURNAL FORESTRY 66 (February 1968): 128-33.

G. RADIOACTIVITY

Radioactive pollution is associated primarily with the disposal of wastes from nuclear power plants, although some wastes come from the testing of atomic weapons, and the production and use of radioactive material in medicine and other research areas. While there are tight controls on the manufacture, transport, use, and disposal of these elements, measurable amounts are still present in our environment. The increasing use of nuclear power plants pose additional problems, particularly in the area of waste disposal. Several of the following references discuss the problems associated with the disposal of such compounds in the sea.

Radioactive elements, like other compounds, react with the various physical, chemical, and biological components of the environment. These reactions affect their form, distribution, and availability to other organisms. As in other areas, the new approach is to study the interaction of the element with the total environment and not just its effect on a single organism.

1. Abstracts and Bibliographies

Edmundson, Eldon, Jr.; Schultz, Vincent; and Klement, Alfred W., Jr. MARINE RADIOECOLOGY: A SELECTED BIBLIOGRAPHY OF NON-RUSSIAN LITERATURE. Washington, D.C.: U.S. Atomic Energy Commission, Division of Technical Information, 1969. Distributed by National Technical Information Service. 127 p.

> Bibliography arranged alphabetically by author only. Information taken primarily from NUCLEAR SCIENCE ABSTRACTS, 1948-68.

International Atomic Energy Agency. DISPOSAL OF RADIOACTIVE WASTES IN MARINE AND FRESH WATERS. Bibliographical Series, no. 5. Vienna: 1962. 365 p.

> Over two thousand references to topics ranging from general information to the sources and nature of radioactive wastes, their treatment, storage, disposal, monitoring, legal, and economic aspects published between 1955-61.

Klement, Alfred W., Jr., and Schultz, Vincent. TERRESTRIAL AND FRESH-WATER RADIOECOLOGY: A SELECTED BIBLIOGRAPHY. Washington, D.C.: National Technical Information Service, 1962-72.

Includes first volume issued in 1962 and eight supplements, the last one published in 1972. Contains bibliographic citations (no index) of articles which report the results of studies on the effects of radiation on all organisms except laboratory strains of animals. Most references from NUCLEAR SCIENCE ABSTRACTS.

Lattimer, John M., ed. ESTUARINE RADIOECOLOGY: A BIBLIOGRAPHY OF REPORT LITERATURE. Washington, D.C.: U.S. Department of the Interior, Office of Library Services, 1970. 52 p.

Bibliographic citations of technical reports published from 1960 through June 1969 concerned with accumulation, cycling, and effects of radioactive wastes on estuarine biota.

Reichle, D.E.; Nelson, D.J.; and Dunaway, P.B. BIOLOGICAL CONCENTRATION AND TURNOVER OF RADIONUCLIDES IN FOOD CHAINS: A BIBLIOGRAPHY. Oak Ridge, Tenn.: Oak Ridge National Laboratory, 1971. Distributed by National Technical Information Service. viii, 42 p.

Bibliographic citations of articles concerned with the consequences of acute or chronic contamination. Cited literature does not date beyond 1969.

[U.S. Department of Interior.] Water Resources Scientific Information Center. STRONTIUM IN WATER: A BIBLIOGRAPHY. Bibliography Series WRSIC 71-201. Washington, D.C.: 1971. Distributed by National Technical Information Service. 80 p.

Citations from SELECTED WATER RESOURCES ABSTRACTS through 15 June 1971 covering the physical and biological aspects of strontium in water.

2. References

International Atomic Energy Agency. DISPOSAL OF RADIOACTIVE WASTES INTO FRESH WATER. Vienna: 1963. Distributed by National Agency for International Publications, New York. 99 p.

General discussion is first presented on the fate of radioactive wastes in freshwater systems, control and monitoring requirements, and legal and administrative principles, then the naturally occurring radionucleotides, and those produced by man are listed with data on radioactivity in selected streams and stream organisms.

______. RADIOACTIVE-WASTE DISPOSAL INTO THE SEA. Vienna: 1961. Distributed by National Agency for International Publications, New York. 174 p.

Contains a good general discussion on the problem of radioactive wastes, their formation and disposal, as well as their effect on man. Includes extensive appendixes with data on amounts of radioactive materials disposed of, types, and sampling and analytical techniques.

National Academy of Sciences. THE EFFECTS OF ATOMIC RADIATION ON OCEANOGRAPHY AND FISHERIES. Washington, D.C.: 1957. ix, 137 p.

> Collection of articles on the properties of radioactive wastes, their disposal and the effect of radiation on aquatic organisms. Technical.

______. RADIOACTIVITY IN THE MARINE ENVIRONMENT. Washington, D.C.: 1971. ix, 272 p. Bibliography.

> Good source for the distribution of radionucleotides in the marine environment, both physical and biological. Considers the physical factors affecting movement of radionucleotides in the oceans and their effect on marine organisms and man.

Polikarpov, G.G. RADIOECOLOGY OF AQUATIC ORGANISMS. Translated by Vincent Schultz and Alfred W. Klement, Jr. New York: Reinhold Publishing, 1966. xxviii, 314 p.

> Focuses attention on Russian activity in the field of marine radioecology with emphasis on the interaction of marine organisms with environmental radioactivity. Presents concentration factors and information on the radiosensitivity of aquatic organisms and their role in the migration of radioactive substances.

Pruter, A.T., and Alverson, Dayten L., eds. THE COLUMBIA RIVER ESTUARY AND ADJACENT OCEAN WATERS. Seattle: University of Washington Press, 1972. 868 p.

> Description of the physical, chemical, and biological aspects of the Columbia River Estuary and measurements of radionucleotides in the physical environment and in the biota. Comprehensive account of the impact of nuclear wastes upon the biota of this river.

Schultz, Vincent, and Klement, Alfred W., Jr., eds. RADIOECOLOGY. New York and Washington, D.C.: Reinhold Publishing, 1963 xvii, 746 p.

> Proceedings of a symposium. Papers deal with cycling of radionucleotides in the environment, both terrestrial and aquatic, and the effects on organisms and plants. Technical.

Schultz, Vincent, and Whicker, F. Ward, comps. ECOLOGICAL ASPECTS OF THE NUCLEAR AGE: SELECTED READINGS IN RADIATION ECOLOGY. Washington, D.C.: Office of Information Services, U.S. Atomic Energy Commission, 1972. 588 p.

> Covers radionuclide cycling and the disposal and effects of radioactive material in terrestrial, freshwater, and marine environments. Contains papers summarizing the effect of radiation on populations, communities, and ecosystems, as well as papers of historical interest.

3. Journal Articles

Baptist, J.P.; Wolfe, D.A.; and Colby, D.R. "Effects of Chronic Gamma Radiation on the Growth and Survival of Juvenile Clams (Mercenaria Mercenaria) and Scallop (Argopecten Irradians)." HEALTH PHYSICS 30 (January 1976): 79-83.

Berg, Gerald G., and Ginsberg, E. "Partition of Ruthenium-106 between the Fresh Water Environment and Crayfish." HEALTH PHYSICS 30 (April 1976): 329-38.

Brisbin, I.L., Jr., et al. "Patterns of Radiocesium in the Sediments of a Stream Channel Contaminated by Production Reactor Effluents." HEALTH PHYSICS 27 (July 1974): 19-27.

Cranmore, G., and Harrison, F.L. "Loss of ^{137}Cs and ^{60}Co from the Oyster Crassostrea Gigas." HEALTH PHYSICS 28 (April 1975): 319-33.

Daly, J.C.; Goodyear, S.; Paperiello, C.J.; and Matuszek, J.M. "Iodine-129 Levels in Milk and Water Near Fuel Reprocessing Plant." HEALTH PHYSICS 26 (April 1974): 333-42.

Englehart, R.W. "An Environmental Analysis of Nuclear Merchant Ship Operation." TRANSACTIONS AMERICAN NUCLEAR SOCIETY 22 (November 1975): 105-6.

Eyman, L.D., and Kevern, N.R. "Cesium-137 and Stable Cesium in a Hypereutrophic Lake." HEALTH PHYSICS 28 (May 1975): 549-55.

Griffin, L.L.; Myers, J.K.; Carlyle, D.C.; and Suttle, A.D., Jr. "Occurrence of Strontium-90 in Living Coral from the Northern Gulf of Mexico." HEALTH PHYSICS 28 (June 1975): 693-97.

Hakonson, T.E., and Whicker, F.W. "Cesium Kinetics in a Montana Lake Ecosystem." HEALTH PHYSICS 28 (June 1975): 699-706.

Martin, J.J., and Boissonnett, B. "Radiological Problems of Liquid Waste Disposal into the Sea." TRANSACTIONS AMERICAN NUCLEAR SOCIETY 20 (April 1975): 423-26.

Menzel, R.G. "Land Surface Erosion and Rainfall as Sources of Strontium-90 in Streams." JOURNAL OF ENVIRONMENTAL QUALITY 3 (July-September 1974): 219-23.

Nevissi, A., and Schell, W.R. "^{210}Po and ^{239}Pu, ^{240}Pu in Biological and Water Samples from the Bikini and Eniwetok Atolls." NATURE 255 (22 May 1975): 321-23.

Yousef, Y.A.; Padden, T.J.; and Gloyna, E.F. "Diurnal Changes in Radionuclides Uptake by Phytoplankton in Small Scale Ecosystems." WATER RESEARCH 9 (February 1975): 181-87.

H. NUISANCE ORGANISMS

Nuisance organisms are considered to be any plant or animal which reduces or eliminates desirable uses of water such as drinking, swimming, boating, or even fishing. Fungi, bacteria, and viruses are also included under this definition. Plant nuisances such as algae are associated with eutrophication, and the reader should consult that section, p. 125, for additional references. Also chapter 17, "Treatment and Management of Water and Wastewater--Including Water Quality Criteria" (p. 149), has references relating to the presence of pathogens in drinking water.

1. Abstracts and Bibliographies

ALGAE ABSTRACTS. 2 vols. New York: IFI/Plenum Publishing, 1973.

> See Part 1, Abstract section under "Biology." Volume 1 covers to 1969, and volume 2 from 1970-72.

Nova Scotia Research Foundation. SELECTED BIBLIOGRAPHY ON ALGAE. 14 vols. Nova Scotia: 1953-73. Author index.

> Bibliographic citations on a wide variety of topics including noxious algae, algalcides, pollution, and pollution control, also physiology, nutrition, and ecology.

2. References

Berg, Gerald [G.], ed. TRANSMISSION OF VIRUSES BY THE WATER ROUTE. New York: Wiley-Interscience, 1967. xviii, 484 p.

> Based on a symposium concerned with viruses in water concentrating on analysis and detection devices.

Cooke, William B. A LABORATORY GUIDE TO FUNGI IN POLLUTED WATERS, SEWAGE AND SEWAGE TREATMENT SYSTEMS: THEIR IDENTIFICATION AND CULTURE. Cincinnati: U.S. Department of Health, Education and Welfare, Public Health Service, 1963. vii, 132 p. Glossary.

> Primarily a key to the fungi of polluted water. Includes section on how to sample, preparation of media, and isolation of fungi. Provides notes on biology of fungi.

______, ed. OUR MOULDY EARTH: A STUDY IN THE FUNGI OF OUR ENVIRONMENT WITH EMPHASIS ON WATER. Cincinnati: U.S. Federal Water Pollution Control Administration, Robert A. Taft Water Research Center, 1970. vii, 533 p.

> Compilation of thirty-nine papers dealing with the various aspects of the biology of fungus. Provides a literature survey as well as techniques for growing fungi and their uses in environmental sanitation and as indicator organisms.

Finstein, Melvin S. POLLUTION MICROBIOLOGY: A LABORATORY MANUAL. New York: Marcell Dekker, 1972. x, 168 p.

> Introduction to the study of pollution microbiology with emphasis on basic concepts underlying microbial aspects of pollution. Contents include: some microorganisms in clean and polluted waters, sanitary bacteriology, carbonaceous pollutants, and nitrogen, phosphorus, sulfur, and iron as pollutants.

Fitzgerald, George P. NUTRIENT SOURCES FOR ALGAE AND THEIR CONTROL. Water Pollution Control Research Series 16010 EHR. Washington, D.C.: U.S. Environmental Protection Agency, Water Quality Office, 1971. Distributed by U.S. Government Printing Office. 77 p.

> Report of a study which evaluated the relationship between algae and their nutrient sources. Study involved a critical evaluation of bioassay techniques, and experiments monitoring the growth and enzymatic activity of algae as related to nutrient supply.

Geldreich, E.E. "Water-Borne Pathogens." In WATER POLLUTION MICROBIOLOGY, edited by Ralph Mitchell, pp. 207-41. New York: Wiley-Interscience, 1972.

> Excellent overview of waterborne pathogens and their public health significance. Covers the common pathogens, as well as discussion on the major source of such pathogens.

Heukelekian, H., and Dondero, N., eds. PRINCIPLES AND APPLICATIONS OF AQUATIC MICROBIOLOGY. New York: John Wiley and Sons, 1964. xii, 452 p.

> Proceedings of a conference which covered the broad aspects of microbiology including microbiology of pesticides and related hydrocarbons, planktonic, and benthic bacteria of lakes, river bacteriology, and the role of bacteria in the cycling of various inorganic compounds such as phosphorus.

Holm, L.G.; Weldon, L.W.; and Blackburn, R.D. "Aquatic Weeds." In MAN'S IMPACT ON ENVIRONMENT, edited by T.R. Detwyler, pp. 246-65. New York: McGraw-Hill, 1971.

Discusses the various aquatic weeds and the problems they have created, as well as chemical and biological control.

Jackson, Daniel F., ed. ALGAE, MAN, AND THE ENVIRONMENT. Syracuse, N.Y.: Syracuse University Press, 1968. viii, 554 p.

Proceedings of a symposium which discussed various aspects of algal growth, requirements, and control. Contents include microorganic and microinorganic requirements, phosphate metabolism, the influence of light and temperature on algae, and effect of type of wastewater treatment on algal growth.

Kerr, Pat C.; Paris, Doris F.; and Brockway, D.L. THE INTERRELATION OF CARBON AND PHOSPHORUS IN REGULATING HETEROTROPHIC AND AUTOTROPHIC POPULATIONS IN AQUATIC ECOSYSTEMS. Water Pollution Control Research Series. Washington, D.C.: Federal Water Quality Administration, 1970. v, 53 p.

Laboratory and field investigations were conducted on the fate and cycling of carbon and phosphorus in selected aquatic ecosystems.

McKinney, Ross E. MICROBIOLOGY FOR SANITARY ENGINEERS. New York: McGraw-Hill, 1962. 293 p.

Examines the fundamental biology and biochemistry of microorganisms important to sanitary engineers. Designed as a textbook.

Mitchell, Ralph, ed. WATER POLLUTION MICROBIOLOGY. New York: Wiley-Interscience, 1972. x, 416 p.

Discusses effects of pollutants on microorganisms, microorganisms associated with pollution and the use of microorganisms for pollution control. Fairly technical.

Rodina, A.G. METHODS IN AQUATIC MICROBIOLOGY. Translated by Rita R. Colwell and Michael S. Zambruski. Baltimore: University Park Press, 1972. xxiii, 461 p.

Topics covered include methods of sampling and culture for various microorganisms, as well as methods for studying the various cycles of inorganic compounds. Techniques for bacteriological analysis of water for public health are provided.

Taft, Clarence E. WATER AND ALGAE: WORLD PROBLEMS. Chicago: Educational Publishers, 1965. xx, 236 p.

Attempts to illustrate the extent to which algae in water is associated with human activities. Includes methods for determination and control of alga problems.

Tryon, Clarence A., Jr., and Hartman, R.T., eds. THE ECOLOGY OF ALGAE. Pittsburgh: Pymatuning Laboratory of Field Biology, University of Pittsburgh, 1960. v, 96 p.

Proceedings of a symposium which covered the many facets of algae ecology including their distribution, organic production, and disturbances resulting from their presence.

Washington, University of, and U.S. Federal Water Pollution Control Administration. ENVIRONMENTAL REQUIREMENTS OF BLUE-GREEN ALGAE. Corvallis, Oreg.: Pacific Northwest Water Laboratory, 1967. v, 111 p.

Collection of symposium papers, generally discussion--type, on the nutritional requirements and physiology of bluegreen algae. Topics include nitrogen nutrition, environmental requirements, and problems in laboratory culture.

Water Quality Conference. VIRUS AND WATER QUALITY: OCCURRENCE AND CONTROL. University of Illinois Bulletin, vol. 69, no. 1 Urbana: University of Illinois, 1971. 224 p.

Collection of papers concerning the detection, removal, and occurrence of viruses in water.

Wood, E.J. Ferguson. MICROBIOLOGY OF OCEANS AND ESTUARIES. Elsevier Oceanography Series 3. New York: Elsevier Publishing, 1967. xi, 319 p.

Covers the various microorganisms of the oceans and estuaries including bacteria and algae and their role in the physical and chemical phenomena of the seas.

3. Journal Articles

Foster, D.H.; Hanes, N.B.; and Lord, S.M., Jr. "A Critical Examination of Bathing Water Quality Standards." JOURNAL WATER POLLUTION CONTROL FEDERATION 43 (November 1971): 2229-41.

Geldreich, E.E. "Applying Bacteriological Parameters to Recreational Water Quality." JOURNAL AMERICAN WATER WORKS ASSOCIATION 62 (February 1970): 113-20.

Gerba, C.P., and Schaiberger, G.E. "Effect of Particulates on Virus Survival in Seawater." JOURNAL WATER POLLUTION CONTROL FEDERATION 47 (January 1975): 93-103.

Gerba, C.P.; Wallis, C.; and Melnick, J.L. "Viruses in Water: the Problem, Some Solutions." ENVIRONMENTAL SCIENCE AND TECHNOLOGY 9 (December 1975): 1122-26.

Novak, J.T.; Goodman, A.S.; and King, D.L. "Aquatic-Weed Decay and Color Production." JOURNAL AMERICAN WATER WORKS ASSOCIATION 67 (March 1975): 134-39.

I. EUTROPHICATION AND ORGANIC POLLUTANTS

The progressive enrichment of waters, with the associated production of algae, increased productivity, and other biotic changes is referred to as eutrophication. It has been and continues to be an important problem of inland waters. Although widely studied, the factors responsible for eutrophication are still not fully understood and workers do not agree on which factors are the most important. The reader is encouraged to read Likens, SYMPOSIUM ON NUTRIENTS AND EUTROPHICATION: THE LIMITING NUTRIENT CONTROVERSY, p. 127, for an excellent overview of the role of nutrients in eutrophication. The basic concepts of eutrophication are presented in the journal articles by Sawyer "ABCs of Cultural Eutrophication and Its Control," parts 1 and 2, p. 131. The article by Hutchinson, p. 127, presents an excellent account of the historical development of scientific thought on this problem.

1. Abstracts and Bibliographies

EUTROPHICATION. Madison, Wis.: Water Resources Information Program. University of Wisconsin, 1972-- . Bimonthly with annual cumulation index.

> Summarizes papers which cover all aspects of eutrophication, popular articles included. Addresses of authors given for reprint requests. Reprints also available from the publisher. Annual index titled: PERMUTED INDEX TO CURRENT EUTROPHICATION LITERATURE.

Mackenthun, Kenneth M. NITROGEN AND PHOSPHORUS IN WATER: AN ANNOTATED BIBLIOGRAPHY OF THEIR BIOLOGICAL EFFECTS. Washington, D.C.: U.S. Department of Health, Education and Welfare, Public Health Service, 1965. Distributed by U.S. Government Printing Office. xxviii, 111 p. Glossary.

> Bibliography presents the results from papers, reports, and other publications (prior to 1965) which deal with nutrients in water, especially nitrogen and phosphorus.

[U.S. Department of Interior.] Water Resources Scientific Information Center. PHOSPHORUS REMOVAL: A BIBLIOGRAPHY. 2 vols. Bibliography Series WRSIC 73-208. Washington, D.C.: 1973. Distributed by National Technical Information Service.

> Citations on phosphorus removal from SELECTED WATER RESOURCES ABSTRACTS up to 15 February 1973.

2. References

Allen, Herbert E. NUTRIENTS IN NATURAL WATERS. New York: Wiley-Interscience, 1972. x, 457 p.

> Coverage includes analysis and bioassay techniques for nutrient measurement; the physical, chemical, and biological processes involved in nutrient recycling; waste treatment aspects and government policies. Also covers the atmospheric contribution of nutrients to the aquatic system.

Brown, Randall L., and Tunzi, Milton G., eds. SEMINAR ON EUTROPHICATION AND BIOSTIMULATION. Sacramento: California Department of Water Resources, 1972. 229 p.

> Contents include: effect of inorganic carbon on eutrophication, role of nitrogen in biostimulation and eutrophication, phosphorus and eutrophication, sediment-water nutrient interchange, effects of light and temperature on algal growth, role of zooplankton in the process of eutrophication, modeling phytoplankton populations, and algal bioassays.

Eppley, Richard. EUTROPHICATION IN COASTAL WATERS: NITROGEN AS A CONTROLLING FACTOR. Water Pollution Control Research Series 16010 EHC. Washington, D.C.: U.S. Environmental Protection Agency, Office of Research and Monitoring, 1971. Distributed by Government Printing Office. ix, 67 p.

> Examination of the role of coastal sewage outfalls in eutrophication, also examined utilization and assmiliation of nitrogen by phytoplankton.

Faust, Samuel D.J., and Hunter, J.V., eds. ORGANIC COMPOUNDS IN AQUATIC ENVIRONMENTS. New York: Marcel Dekker, 1971. xxii, 638 p.

> Papers presented at a conference which dealt primarily with the chemistry of organic materials in water and their degradation. Most papers cover the many possible chemical reactions and the effect on the aquatic environment, such as adsorption thermodynamics for organics on surfaces, metalorganic interactions in the marine environment, photodecomposition of organic pesticides, energetics, and bacterial growth. Technical.

Goldman, Charles R.; Moshiri, Gerald; and deAmezaga, Evelyne. "Synoptic Study of Accelerated Eutrophication in Lake Tahoe--An Alpine Lake." In INTERNATIONAL SYMPOSIUM ON WATER POLLUTION CONTROL IN COLD CLIMATES, edited by R. Sage Murphy and David Nyquist, pp. 1-21. Washington, D.C.: U.S. Environmental Protection Agency, Office of Research and Monitoring, 1971.

Results of a study on the primary productivity of Lake Tahoe and its relation to eutrophication. Primary producers were viewed as a more sensitive indicator of increased fertility than chemical parameters.

Goldman, Joel C.; Porcella, Donald B.; and Middlebrooks, E. Joe. "The Effect of Carbon on Algal Growth--Its Relationship to Eutrophication." WATER RESEARCH 6 (June 1972): 637-79.

Discussion of the carbon cycle in aquatic systems including its effect on algal growth. Article is a review of the literature up to 1971.

Hutchinson, G. Evelyn. "Eutrophication." AMERICAN SCIENTIST 61 (May-June 1973): 269-79.

Good review of the various aspects of eutrophication. Topics included are an historical perspective, general lake ecology, the nutrition of phytoplankton, and limiting factors.

International Association on Water Pollution Research. "Phosphorus in Freshwater and the Marine Environment." WATER RESEARCH 7 (January-February 1973): 1-342.

Papers given at a conference in 1972 on the presence of phosphorus in aquatic systems, papers cover methods of evaluation and removal, the significance of manmade sources of phosphorus, the role of phosphorus in eutrophication, and sources of phosphorus from industry and agriculture.

Lake Tahoe Area Council. EUTROPHICATION OF SURFACE WATERS--LAKE TAHOE: FOURTH PROGRESS REPORT. Water Pollution Control Research Series, no. 16011 DSW. Washington, D.C.: U.S. Environmental Protection Agency, Water Quality Office, 1971. Distributed by U.S. Government Printing Office. 1971. xi, 154 p.

Report of a study which examined the nutrient and chemical constituents of the lake and the potential for algal growth stimulation. Included is a survey of nutrients and other chemical constituents, along with results from bioassays and other analyses.

Likens, G.E., ed. SYMPOSIUM ON NUTRIENTS AND EUTROPHICATION: THE LIMITING NUTRIENT CONTROVERSY. Special Symposia. Vol. 1. Lawrence, Kans.: Allen Press, 1972. 328 p.

Proceedings of a symposium in 1971 which explored the current thinking relative to the importance of various regulating or limiting nutrients in the eutrophication of aquatic ecosystems, specifically nutrient availability. Major focus was on carbon and phosphorus and their role in regulating eutrophication.

Lund, J.W.G. "Eutrophication." PROCEEDINGS ROYAL SOCIETY OF LONDON. Series B 180 (21 March 1972): 371-82.

Comparison of eutrophic conditions in two aquatic systems and the affect of water chemistry on the eutrophication processes. Most emphasis is on the effect of phosphate compounds on aquatic life.

Middlebrooks, E. Joe; Falkenborg, Donna H.; and Maloney, Thomas E. MODELING THE EUTROPHICATION PROCESS. Ann Arbor, Mich.: Ann Arbor Publishers, 1973. viii, 228 p.

Proceedings of a workshop on the concepts applicable to modeling ecological or biological processes associated with eutrophication. Topics include modeling algal growth dynamics, aspects of phosphorus dynamics, role of thermocline migration, and multinutrient dynamics model. Technical.

Middlebrooks, E. J[oe].; Maloney, T.E.; Powers, D.F.; and Kaack, L.M., eds. PROCEEDINGS OF EUTROPHICATION AND BIOSTIMULATION ASSESSMENT WORKSHOP. Berkeley and Los Angeles: University of California, 1969. 281 p.

Principal areas under scrutiny at this conference on eutrophication were: understanding the problem, ecological control, mechanical control, chemical control, and biological control. Papers were presented on assay procedures, algal physiology, limnological investigations, and eutrophication control and applications.

Moore, Charles A., and Silver, Marshall L. THE ROLE OF SEDIMENTS IN EUTROPHICATION--A PRELIMINARY STUDY. Research Report, no. 50. Urbana: University of Illinois, Water Resources Center, 1972. Distributed by National Technical Information Service. 89 p.

Report is divided into two sections: first part is an extensive literature survey detailing the distribution of nutrients in water and sediments, and the mechanisms for describing phosphate mobility in aquatic systems, second part presents laboratory techniques for the study of the transport of phosphates in soil.

Morton, Stephen D.; Derse, Philip H.; and Sernau, Russell C. THE CARBON DIOXIDE SYSTEM AND EUTROPHICATION. Water Pollution Control Research Series 16010 DXV. Washington, D.C.: U.S. Environmental Protection Agency, Office of Research and Monitoring, 1971. vii, 72 p.

Report of a study which examined the growth rates of algae with respect to carbon availability, and the feasibility of controlling carbon concentrations in natural bodies of water.

Murphy, Thomas J. SOURCES OF PHOSPHORUS INPUTS FROM THE ATMOSPHERE AND THEIR SIGNIFICANCE TO OLIGOTROPHIC LAKES. Water Resources Center Research Report, no. 92. Urbana: University of Illinois, 1974. iii, 45 p.

Final report of a study which analyzed precipitation in the Chicago area for phosphorus and determined if atmospheric sources made an important contribution to the total phosphorus in Lake Michigan. Report includes data on the composition of the precipitation and a determination of the sources of phosphorus entering Lake Michigan.

National Academy of Sciences. ACCUMULATION OF NITRATE. Report no. 2038. Washington, D.C.: 1972. 106 p.

Report examines the various problems associated with the accumulation of nitrate, nitrogen, and related nitrogenous compounds in the environment and recommends courses of action to mitigate these problems. Chapter on eutrophication includes sections on: analytical techniques, monitoring and surveillance, prevention and control, anticipated consequences and costs of control, criteria and standards, and regulatory and legislative needs.

______. EUTROPHICATION: CAUSES, CONSEQUENCES, CORRECTIVES. Washington, D.C.: 1969. vii, 661 p.

Proceedings of a symposium which discussed many facets of the eutrophication process, including some geographical concepts, methods for its detection and measurement, preventive and corrective measures, and recommendations for management.

Organization for Economic Cooperation and Development. EUTROPHICATION IN LARGE LAKES AND IMPOUNDMENTS. London: H.M.S.O., 1970. 560 p.

Compilation of case studies--European--on eutrophication. Each study includes data on the biological aspects, chemical features, methods analysis, and management techniques.

Putnam, H.D.; Brezonik, P.L.; and Shannon, E.E. EUTROPHICATION FACTORS IN NORTH CENTRAL FLORIDA LAKES. Washington, D.C.: U.S. Environmental Protection Agency, Office of Research and Monitoring, 1972. xi, 141 p.

Case study detailing the physical, chemical, and biological characteristics of some Florida lakes and their level of eutrophication. Includes a study of algal response to nutrient addition along with their chlorophyll levels.

Stewart, Kenton M., and Rohlich, Gerald A. EUTROPHICATION--A REVIEW. Sacramento: California State Water Quality Control Board, 1967. 188 p.

Review and evaluation of the scientific and descriptive work published on eutrophication. Includes lake classification, nutrients, analysis, and management procedures. Contains an extensive bibliography of the literature up to 1967. Glossary.

Vollenweider, Richard A., ed. A MANUAL ON METHODS FOR MEASURING PRIMARY PRODUCTION IN AQUATIC ENVIRONMENTS INCLUDING A CHAPTER ON BACTERIA. International Biological Programme Handbook, no. 12. Oxford: Blackwell Scientific Publications, 1969. xvi, 213 p.

Manual of methods written by experts, includes sampling techniques, methods for measuring production rates, and measurements on isolated and nonisolated communities.

3. Journal Articles

Beeton, A.M. "Eutrophication of the St. Lawrence--Great Lakes." In MAN'S IMPACT ON ENVIRONMENT, edited by T.R. Detwyler, pp. 233-45. New York: McGraw-Hill, 1971.

Crossland, J., and McCaull, J. "Overfed." ENVIRONMENT 14 (September 1973): 30-37.

Dunstan, W.M. "Problems of Measuring and Predicting Influence of Effluents on Marine Phytoplankton." ENVIRONMENTAL SCIENCE AND TECHNOLOGY 9 (July 1975): 635-38.

Duursma, E.K., and Marchand, M. "Aspects of Organic Marine Pollution." OCEANOGRAPHY AND MARINE BIOLOGY: AN ANNUAL REVIEW 12 (1974): 315-431.

Fjerdingstad, E. "Microbial Criteria of Environmental Qualities." ANNUAL REVIEW OF MICROBIOLOGY 25 (1971): 563-82.

Foehrenbach, J. "Eutrophication." JOURNAL WATER POLLUTION CONTROL FEDERATION 44 (June 1972): 1150-59.

Fogg, G.E. "Nitrogen Fixation in Lakes." PLANT AND SOIL. Special volume (1971): 393-401.

Imboden, D.M. "Phosphorus Model of Lake Eutrophication." LIMNOLOGY AND OCEANOGRAPHY 19 (March 1974): 297-304.

Morton, S[tephen]. D., and Lee, T.H. "Algal Blooms--Possible Effects on Iron." ENVIRONMENTAL SCIENCE AND TECHNOLOGY 8 (July 1974): 673-74.

O'Brien, W.J. "Changes in Plankton Species Composition and Diversity in a Controlled Nutrient Enrichment Study." TRANSACTIONS OF THE AMERICAN MICROSCOPICAL SOCIETY 91 (January 1972): 77-80.

Regier, H.A., and Hartman, W.L. "Lake Erie's Fish Community: 150 Years of Cultural Stresses." SCIENCE 180 (22 June 1973): 1248-55.

Sawyer, C[lair]. N. "ABCs of Cultural Eutrophication and Its Control: Part 1--Cultural Changes." WATER AND SEWAGE WORKS 118 (September 1971): 278-81.

______. "ABCs of Cultural Eutrophication and Its Control: Part 2--Waste Waters." WATER AND SEWAGE WORKS 118 (October 1971): 322-27.

Chapter 16
POLLUTION SOURCES

A. INDUSTRIAL POLLUTION

Many pollutants enter the environment mixed with other compounds and are, therefore, most easily classified by their origin. Industrial pollution encompasses wastes from such diverse sources as power generation, lumbering and paper processing, mining, the manufacture of chemicals and textiles, and the steel industry. Wastes from the processing of agricultural products are also included in this section.

The following references are concerned primarily with characterizing the effluent and describing treatment methods. References on specific industrial pollutants can be found in the journal articles or in the previous sections under heavy metals, oil, radioactivity, thermal pollution, and so forth.

1. Abstracts and Bibliographies

[U.S. Department of Interior.] Water Resources Scientific Information Center. ACID MINE WATER: A BIBLIOGRAPHY. Bibliography Series 75-202. Washington, D.C.: 1975. Distributed by National Technical Information Service. iv, 564 p.

> Information base: SELECTED WATER RESOURCES ABSTRACTS through 15 November 1974 covers all topics relating to acid mine water.

______. PCB IN WATER: A BIBLIOGRAPHY. Bibliography Series WRSIC 73-201. Washington, D.C.: 1973. Distributed by National Technical Information Service. 144 p.

> Abstracts from SELECTED WATER RESOURCES ABSTRACTS (up to 15 December 1972) pertaining to PCBs in water. This volume supersedes WRSIC 72-201 published in 1972.

2. References

ABT Associates, Inc. INCENTIVES TO INDUSTRY FOR WATER POLLUTION CONTROL: POLICY CONSIDERATIONS. Washington, D.C.: Federal Water Pollution Control Administration, 1967. 119 p.

> Limited chiefly to the manufacturing industry, this report examines the current situation with regard to control of water pollution. Fiscal incentives, stream hydrology, implications in pollution control, and proposals for new policies, such as effluent charges, are a few of the topics discussed.

American Society for Testing and Materials. MANUAL ON WATER. 3d ed. A.S.T.M. Special Technical Publication, no. 442. Philadelphia: 1969. x, 360 p.

> Aimed primarily at industry this reference covers water sources and supplies, water use by industry, and analysis of industrial wastewater. Not detailed.

Barcus, W. Herman, et al. INDUSTRIAL OILY WASTE CONTROL. New York: American Petroleum Institute, 1969. ix, 144 p.

> Discusses sources, analysis, and control of oily wastes from industry, but not oil spills. Provides some statistics on sources and amounts of oily waste discharge.

Berkowitz, David A., and Squires, Arthur M., eds. POWER GENERATION AND ENVIRONMENTAL CHANGE. Cambridge: MIT Press, 1971. xxiii, 440 p.

> Symposium of the Committee on Environmental Alteration, American Association for the Advancement of Science. Includes a wide variety of papers dealing with power generation, several of which are related to water pollution, such as the effects of hydroelectric dams and the impact of waste heat on aquatic organisms. Most of papers are of a general nature.

Berkowitz, J.B. ASSESSING WATER POLLUTION POTENTIAL OF MANUFACTURED PRODUCTS. 4 vols. U.S. Environmental Protection Technology Series, R2 73-179 a,b,c,d. Washington, D.C.: U.S. Environmental Protection Agency, 1973.

> Catalog of manufactured products which could, during their normal use or disposal, result in water pollution. The catalog is in three sections: section 1 summarizes the pollution potential of each listed product, section 2 provides data on chemical composition of each product and pollution effects, section 3 provides listing of chemicals and their standard industrial classification (SIC) codes.

Council on Economic Priorities. PAPER PROFITS: POLLUTION IN THE PULP AND PAPER INDUSTRY. Washington, D.C.: 1971.

> Report contains statistics on over 130 pulp and paper mills in the United States. Information includes amount of water used, (biochemical oxygen demand) of wastes and treatment facilities in operation or needed. Good source for statistics.

Gehm, Harry. STATE-OF-THE-ART REVIEW OF PULP AND PAPER WASTE TREATMENT. Environmental Protection Technical Series, EPA - R2-73-184. Washington, D.C.: U.S. Environmental Protection Agency, 1973. Distributed by U.S. Government Printing Office. xiii, 252 p.

> Report reviews the major production processes, water quality problems and common analytical methods. Discusses the problems of control and treatment.

Gurnham, Charles F., ed. INDUSTRIAL WASTEWATER CONTROL. New York: Academic Press, 1965. x, 476 p.

> Each chapter describes a separate industry indicating its wastewater characteristics, methods of treatment and disposal, and significant trends in reducing wastewater output. Industries range from the various food products to mining and manufactured products.

Hansmann, Eugene W., and Phinney, Harry K. "Effects of Logging on Periphyton in Coastal Streams of Oregon." ECOLOGY 54 (Winter 1973): 194-99.

> Results of a study which observed changes in stream algal flora during a logging study of small watersheds. Three areas were studied: one of which was clearcut, a second was patchcut leaving a buffer strip of vegetation along the stream channel, and a third was kept intact as a control. Analysis indicates that clearcutting had an immediate effect on species composition and increases primary production.

International Atomic Energy Agency. SYMPOSIUM ON ENVIRONMENTAL ASPECTS OF NUCLEAR POWER STATIONS. New York: 1971. 970 p.

> Papers given at a symposium which discussed future needs for electrical power, standards for control of effluents, siting of power stations, waste-heat discharge, radioactivity of emissions, and aesthetic requirements.

Likens, G.E.; Bormann, F.H.; Johnson, W.M.; Fisher, D.W.; and Pierce, R.S. "Effects of Forest Cutting and Herbicide Treatment on Nutrient Budgets in the Hubbard Brook Watershed-Ecosystem." ECOLOGICAL MONOGRAPHS 40 (Winter 1970): 23-47.

> Detailed study of a deforested watershed which examined the

biogeochemical interactions and implications resulting from a large scale change in habitat or vegetation. Study monitored the physical and chemical parameters of the stream and noted significant pollution resulting primarily from a forty-fold increase in nitrate concentration, and an increase in water temperature. Detailed results are given.

Lund, Leonard. INDUSTRY EXPENDITURES FOR WATER POLLUTION ABATEMENT. Conference Board Report, no. 541. New York: Conference Board, 1972. iv, 108 p.

Report covers the growth of capital expenditures in industry from 1965-69, the expenditure and personnel for operation of abatement equipment in 1969 and the volume and characteristics of discharged wastewaters. Report provides basic data on efforts to deal with water pollution problems by selected industries.

Mancy, K[halil].H., and Weber, W.J., Jr. ANALYSIS OF INDUSTRIAL WASTEWATERS. New York: Wiley-Interscience, 1972. 160 p.

Contents of this text include characterization of the wastewaters produced by various industries and presentation of sampling and analytical techniques applicable for the identification and quantification of the components of industrial wastewater. Emphasis is on the design of measurement systems and the theory of analysis.

National Council of the Paper Industry for Air and Stream Improvement. NCASI TECHNICAL BULLETIN. New York: 1957-- . Title varies. Irregular.

Bulletins issued at least monthly; each deals with a separate topic, i.e., STREAM IMPROVEMENT TECHNICAL BULLETIN, no. 275 was a "Review of the Literature on Pulp and Paper Effluent Management." Topics covered are generally related to the operation of the paper industry and their effect on streams.

Nemerow, Nelson L. LIQUID WASTE OF INDUSTRY: THEORIES, PRACTICES AND TREATMENT. Reading, Mass.: Addison-Wesley Publishing, 1971. xiii, 584 p.

Text provides a background in stream pollution, i.e., what it is, how to measure it, and water quality classification. In addition, theories on treating wastewaters are presented. References also includes statistics on the types of wastes produced by many industries. Semitechnical.

Ohio State University Research Foundation. ACID MINE DRAINAGE FORMATION AND ABATEMENT. Water Pollution Control Research Series No. 14010 FPR. Washington, D.C.: U.S. Environmental Protection Agency, Water Quality Office, 1971. 82 p.

Review of current knowledge concerning acid mine drainage formation and abatement procedures.

Russell, F., and Gilson, H.C., eds. "A Discussion on Freshwater and Estuarine Studies of the Effects of Industry." PROCEEDINGS OF THE ROYAL SOCIETY OF LONDON Series B, 180 (March 1972): 363-536.

Collection of articles presented at a discussion session dealing with the various types of water pollution related to industrial production. Topics include thermal pollution effects, radioactivity in freshwater and estuaries, forecasting pollution effects, effects of paper mills on benthos, and effects of oil industry on shore life.

Smith, Ronald W., and Frey, D.G. ACID MINE POLLUTION EFFECTS ON LAKE BIOLOGY. Water Pollution Research Series. Washington, D.C.: U.S. Environmental Protection Agency, 1971. 133 p.

Provides fundamental biological and physical information concerning strip mine lakes to be used as a basis for the sound management of the lakes. Includes biology as well as acid chemistry of the lake. Many statistics.

U.S. Council on Environmental Quality. THE ECONOMIC IMPACT OF POLLUTION CONTROL: A SUMMARY OF RECENT STUDIES. Washington, D.C.: U.S. Government Printing Office, 1972. 332 p.

Contains reprints of the summaries of reports prepared by various contractors on the impact of pollution control costs in such industries as automotive, baking, cement, iron foundry, smelting, and petroleum. Reference puts into perspective studies which were conducted to assess the economic impact of air and water pollution abatement requirements on a number of industries.

3. Journal Articles

Afghan, B.K.; Kulkarni, A.V.; Leung, R.; and Ryan, J.F. "The Determination of Formaldehyde and Related Compounds in Water and Industrial Effleunts." ENVIRONMENTAL LETTERS 7, no. 1 (1974): 53-65.

Ahmed, A.K. "PCBs in the Environment." ENVIRONMENT 18 (March 1976): 6-11.

Bard, C.C.; Murphy, J.J.; Stone, D.L.; and Terhaar, C.J. "Silver in Photoprocessing Effluents." JOURNAL WATER POLLUTION CONTROL FEDERATION 48 (February 1976): 389-94.

Bormann, F.H., and Likens, G.E. "The Watershed Ecosystem Concept and Studies of Nutrient Cycles." In ECOSYSTEM CONCEPTION NATURAL RESOURCES MANAGEMENT, edited by G.M. vanDyne, pp. 49-76. New York: Academic Press, 1969.

Bormann, F.H.; Likens, G.E.; Fisher, D.W.; and Pierce, R.S. "Nutrient Loss Accelerated by Clear-Cutting of a Forest Ecosystem." SCIENCE 159 (23 February 1968): 882-84.

Brook, A.J., and Baker, A.L. "Chlorination at Power Plants: Impact on Phytoplankton Productivity." SCIENCE 176 (30 June 1972): 1414-15.

Brown, George A., and Krygier, J.T. "Effects of Clear-Cutting on Stream Temperature." WATER RESOURCES RESEARCH 6 (August 1970): 1133-39.

Burns, J.W. "Some Effects of Logging and Associated Road Construction on Northern California Streams." TRANSACTIONS AMERICAN FISHERY SOCIETY 101 (January 1972): 1-17.

Carpenter, E.J.; Peck, B.B.; and Anderson, S.J. "Survival of Copepods Passing through a Nuclear Power Station on Northeastern Long Island Sound, USA." MARINE BIOLOGY 24 (February 1974): 49-55.

Eschner, A.R., and Larmoyeux, J. "Logging and Trout: Four Experimental Forest Practices and Their Effect on Water Quality." PROGRESSIVE FISH CULTURIST 25 (April 1963): 59-67.

Gabriel, P.L.; Dias, N.S.; and Nelson-Smith, A. "Temporal Changes in the Plankton of an Industrialized Estuary." ESTUARINE AND COASTAL MARINE SCIENCE 3 (April 1975): 145-51.

Hays, H., and Cormons, G. "Plastic Particles Found in Tern Pellets, on Coastal Beaches and at Factory Sites." MARINE POLLUTION BULLETIN 5 (March 1974): 44-46.

Highland, J. "PCB's in Food." ENVIRONMENT 18 (March 1976): 12-16.

Kay, G.H. "Asbestos in Drinking Water." JOURNAL AMERICAN WATER WORKS ASSOCIATION 66 (September 1974): 513-14.

Martell, J.M.; Rickert, D.A.; and Siegel, F.R. "PCB's in Suburban Watersheds, Reston, Virginia." ENVIRONMENTAL SCIENCE AND TECHNOLOGY 9 (September 1975): 872-74.

Minor, P.S. "Organic Chemical Industry's Waste Waters." ENVIRONMENTAL SCIENCE AND TECHNOLOGY 8 (July 1974): 620-25.

Moore, C.J.; Stevens, G.A.; McErlean, A.J.; and Zion, H.H. "A Sport Fishing Survey in the Vicinity of a Steam Electric Station on the Patuxent Estuary, Maryland." CHESAPEAKE SCIENCE 14 (September 1973): 160-70.

Morris, A.W., and Hamilton, E.I. "Polystyrene Spherules in the Bristol Channel." MARINE POLLUTION BULLETIN 5 (February 1974): 26-27.

Olson, H.L. "Asbestos in Potable-Water Supplies." JOURNAL AMERICAN WATER WORKS ASSOCIATION 66 (September 1974): 515-18.

Paterson, C.G., and Nursall, J.R. 'The Effects of Domestic and Industrial Effluents on a Large Turbulent River." WATER RESEARCH 9 (April 1975): 425-35.

Sanborn, J.R.; Metcalf, R.L.; Yu, C.C.; and Lu, F.Y. "Plasticizers in the Environment.: The Fate of Di-n-octyl Phthalate (DOP) in Two Model Ecosystems and Uptake and Metabolism of DOP by Aquatic Organisms." ARCHIVES OF ENVIRONMENTAL CONTAMINATION AND TOXICOLOGY 3, no. 2 (1975): 244-55.

Schulz, M. "Effect of Aircraft Deicer on Airport Storm Runoff." JOURNAL WATER POLLUTION CONTROL FEDERATION 46 (January 1974): 173-80.

Sharpe, W.E. "Timber Management Influences on Aquatic Ecosystems and Recommendations for Future Research." WATER RESOURCES BULLETIN 11 (June 1975): 546-50.

Shea, K.P. "PCB." ENVIRONMENT 15 (September 1973): 25-28.

Sprules, W.G. "Midsummer Crustacean Zooplankton Communities in Acid-Stressed Lakes." JOURNAL FISHERIES RESEARCH OF CANADA 32 (March 1975): 9.

B. AGRICULTURAL POLLUTION

The types of pollution associated with agriculture include pesticides, eutrophication, pathogens, organic wastes, suspended matter, salts, and nuisance organisms. The reader is directed to those sections for additional information on these pollutants. The reference by Brady, p. 140, is a good overview of the pollution problems arising from agricultural practices, and provides an introduction to the rest of the literature. References on the treatment of agricultural wastes are found both in this section and in chapter 17, "Treatment and Management of Water and Wastewater--Including Water Quality Control," p. 149.

An increasingly popular area of research is the combining of waste treatment and agriculture (see the section on journal articles in chapter 17, p. 149). This involves the application of agricultural or municipal wastes to croplands and the recovery of animal wastes for feed. Another area of research is concerned with the use of plants for waste removal in water treatment, the plants would then be harvested.

Pollution Sources

1. Abstracts and Bibliographies

Miner, J.R. BIBLIOGRAPHY OF LIVESTOCK WASTE MANAGEMENT. Environmental Protection Technology Series, R2-72-101. Washington, D.C.: U.S. Environmental Protection Agency, 1972. 138 p.

Bibliography of the literature relating to animal waste disposal and pollution control. Includes literature on the characterization, storage, transport, treatment, utilization, and disposal of livestock waste.

[U.S. Department of Interior.] Water Resources Scientific Information Center. AGRICULTURAL RUNOFF--A BIBLIOGRAPHY. Bibliography Series, WRSIC 72-204. Washington, D.C.: 1972. Distributed by National Technical Information Service. iii, 248 p.

Abstracts from SELECTED WATER RESOURCES ABSTRACTS for the period 1967 through 15 December 1971. All facets of the influence of agricultural runoff on water quality are covered.

______. IRRIGATION RETURN FLOW--A BIBLIOGRAPHY. Bibliography Series, OWRT/WRSIC 75-209. Washington, D.C.: 1975. Distributed by National Technical Information Service. vi, 207 p.

Abstracts from SELECTED WATER RESOURCES ABSTRACTS through October 1975 relating to the effect of irrigation return flow on water quality, and characterization of the return flow, i.e., its composition and source.

U.S. Public Health Service. AN ANNOTATED BIBLIOGRAPHY ON EFFECTS OF IRRIGATION PRACTICES ON STREAM WATER QUALITY WITH EMPHASIS ON SALINITY AND PESTICIDES. Washington, D.C.: 1963. 37 p.

Includes abstracts of the literature related primarily to salinity from irrigation waters and pesticides and their affect on water quality.

2. References

Brady, Nyle C., ed. AGRICULTURE AND THE QUALITY OF OUR ENVIRONMENT. AAAS Publication, no. 85. Washington, D.C.: American Association for the Advancement of Science Publications, 1967. xv, 460 p.

Section on agriculture and water quality covers the effects of the following on water quality: pesticides, fertilizer nutrients, agricultural processing wastes, sediments, and salinity from irrigation return water. Discusses eutrophication and the economic and administrative problems associated with water pollution. Each area is covered fairly completely.

California. University of. Agricultural Extension Service. POLLUTION CONTROL AND IMPACT UPON AGRICULTURE. Davis, Calif.: 1970.

Papers given at a conference on the various aspects of agriculture and pollution includes both the pollution caused by agricultural practices and the effects of pollution on agriculture. Papers cover all types of pollution. Only one paper is specifically on water pollution.

Cornell University Conference on Agricultural Waste Management. RELATIONSHIP OF AGRICULTURE TO SOIL AND WATER POLLUTION. Ithaca, N.Y.: Cornell University, 1970. 270 p.

Covers a broad range of topics including bacterial indicators in farm surface runoff, effects of runoff from noncultivated rangeland on stream water quality, and water quality of the runoff from land in various stages of agricultural production. Technical.

Doneen, Lloyd D., ed. AGRICULTURAL WASTEWATERS, SYMPOSIUM. Water Resources Center Report, no. 10. Berkeley: University of California, Water Resources Center, 1966. ix, 368 p.

Proceedings of a symposium which discusses the whole range of agricultural practices contributing to formation of waste waters, including irrigation, application of fertilizers and pesticides, and soil erosion from cultivation. Discusses management and treatment methods.

Duffer, William R.; Kreis, R. Douglas; and Harlin, Curtis C., Jr. EFFECT OF FEEDLOT RUNOFF ON WATER QUALITY OF IMPOUNDMENTS. Water Pollution Control Research Series, no. 16080 GGP. Washington, D.C.: U.S. Environmental Protection Agency, Water Quality Office, 1971. Distributed by U.S. Government Printing Office. ix, 53 p.

Results from a study which examined the effect of rainfall-runoff from a beef cattle feedlot on a small impoundment. Both chemical and biological (i.e., community composition) were examined and compared to a control reservoir. There were distinct chemical and biological changes including the death of 90 percent of the game fish.

Law, James P., Jr., and Skogerboe, Gaylord V. "Potential for Controlling Quality of Irrigation Return Flows." JOURNAL OF ENVIRONMENTAL QUALITY 1 (April-June 1972): 140-45.

Potential pollutants related to irrigation return flows such as salinity, nutrients, sediments, and pesticides are reviewed. Suggestions for control measures are given.

Lunin, Jesse. "Agricultural Wastes and Environmental Pollution." In ADVANC IN ENVIRONMENTAL SCIENCE AND TECHNOLOGY, Vol. 2, edited by Ja N. Pitts, Jr., and Robert L. Metcalf, pp. 215-61. New York: Wiley-Intersci 1971.

Review of water pollution problems caused by agriculture, including such potential pollutants as sediments, salts, animal wastes, and agricultural chemicals such as pesticides, also processing wastes. Discusses methods of evaluating agricultural pollution and additional factors affecting pollution.

Middlebrooks, E. Joe. "Animal Wastes, Management and Characterization WATER RESEARCH 8 (October 1974): 697-712.

Reviews the literature to 1973 pertaining to characterization of wastes and runoff and to treatment methods. Contains a lot of statistics.

North Carolina State University. Department of Biological and Agricultura Engineering. ROLE OF ANIMAL WASTES IN AGRICULTURAL LAND RUNO Water Pollution Control Research Series, 13020-DGX 08/71. Washington, D U.S. Environmental Protection Agency, 1971. Distributed by U.S. Governm Printing Office. x, 114 p.

Results of a study on the effects of disposal of animal feedlot wastes on water quality. Report examines pollution from swine, dairy, beef, and poultry production in an attempt to determine the best management methods.

Scalf, Marion R.; Hauser, V.L.; McMillion, L.G.; Dunlap, W.J.; and Keeb J.W. FATE OF DDT AND NITRATE IN GROUND WATER. Washington, D.C U.S. Department of Interior, 1968. ix, 46 p.

Report of a project which examined the fate of DDT and nitrates from agricultural runoff waters which were introduced into an aquifer. Technical.

Willrich, Ted L., and Smith, George E., eds. AGRICULTURAL PRACTICES AND WATER QUALITY. Ames: Iowa State University Press, 1970. xxvii, 415 p.

Individual papers discuss the various aspects of sediments, plant nutrients, pesticides, and animal wastes as water pollutants. Legal and economic aspects also included. Most articles are technical.

3. Journal Articles

Barber, J.C. "Pollution Control in Fertilizer Manufacture." JOURNAL ENVIRONMENTAL QUALITY 4 (January-March 1975): 1-11.

Branson, R.L.; Pratt, P.F.; Rhodes, J.D.; and Oster, J.D. "Water Quality in Irrigated Watersheds." JOURNAL ENVIRONMENTAL QUALITY 4 (January-March 1975): 33-40.

Brink, N. "Water Pollution from Agriculture." JOURNAL WATER POLLUTION CONTROL FEDERATION 47 (April 1975): 789-95.

Filip, D.S., and Middlebrooks, E.J. "Eutrophication Potential of Dairy Cattle Waste Runoff." WATER RESEARCH 10 (January 1976): 89-93.

Filip, D.S.; Middlebrooks, E.J.; and Porcella, D.B. "Chemical and Bio-Stimulation Properties of Cattle Feedlot Runoff." WATER RESEARCH 9 (May-June 1975): 573-79.

Gostick, K.G., and Dermott, W. "Agricultural Aspects of Nitrates in Water." WATER TREATMENT AND EXAMINATION 24, no. 3 (1975): 206-33.

Harms, L.L.; Middaugh, P.; Dornbush, J.N.; and Andersen, J.R. "Agricultural Runoff Pollutes Surface Waters--Part I." WATER AND SEWAGE WORKS 122 (October 1975): 84-85.

Humenik, F.J.; Overcash, M.R.; Diggers, L.B.; and Kriz, G.J. "Cleaning the Animal Farm Environment." ENVIRONMENTAL SCIENCE AND TECHNOLOGY 8 (November 1974): 984-89.

Nicholson, H.P. "The Needs for Water Quality Models on Agricultural Watersheds." JOURNAL ENVIRONMENTAL QUALITY 4 (January-March 1975): 21-23.

Olness, A.; Smith, S.J.; Rhoades, E.D.; and Menzel, R.G. "Nutrient and Sediment Discharge from Agricultural Watersheds in Oklahoma." JOURNAL ENVIRONMENTAL QUALITY 4 (July-September 1975): 331-36.

Shabi, F.H., and Cannon, M.C. "Characteristics and Treatment of Dairy and Meat Effluents." EFFLUENT AND WATER TREATMENT JOURNAL 15 (April 1975): 176-77.

Viets, F.G. "The Environmental Impact of Fertilizers." CRC CRITICAL REVIEWS IN ENVIRONMENTAL CONTROL 5 (September 1975): 423-53.

C. URBAN POLLUTION

Water pollution from urban areas results from the disposal of municipal wastes and runoff from streets and construction projects. Improperly or untreated municipal wastes add nutrients and pathogens to receiving waters, while runoff is a source of oil, heavy metals, suspended matter, pesticides, and salts. Land development contributes to water pollution through soil disturbances which result in siltation of adjacent waters. In addition, waterfront property is fashionable and many developments channelize streams, fill in marsh and shoreline areas of lakes, or otherwise modify a water course to suit their needs. Improperly done, such modifications can disturb or destroy the aquatic habitat.

Most of the above pollutants have been covered in other sections and the reader should refer to them for fuller coverage. The references in this section are concerned primarily with the impact of urban runoff. The journal articles update the book literature and are indicative of the journals containing articles on urban pollution.

1. Abstracts and Bibliographies

[U.S. Department of Interior.] Water Resources Scientific Information Center. URBANIZATION AND SEDIMENTATION: A BIBLIOGRAPHY. 2 vols. Bibliography Series, WRSIC 71-203 and OWRT/WRSIC 75-207. Washington, D.C.: 1971, 1975. Distributed by National Technical Information Service.

> Abstracts of the literature concerned with urban development, sedimentation, water resources planning, land use, and water quality. Volume 1 covers the period October 1968 through September 1971. Volume 2 covers October 1971 through October 1975. Abstracts for both volumes from SELECTED WATER RESOURCES ABSTRACTS.

_____. URBAN WATER PLANNING: A BIBLIOGRAPHY. Bibliography Series, WRSIC 72-215. Washington, D.C.: 1972. Distributed by National Technical Information Service. 373 p.

> Abstracts from SELECTED WATER RESOURCES ABSTRACTS through 15 May 1972, covering urban planning and urban water planning.

U.S. Environmental Protection Agency. Office of Research and Monitoring. SELECTED URBAN STORM WATER RUNOFF ABSTRACTS. 8 vols. Environmental Protection Technology Series. Washington, D.C.: 1969-72.

> Series of eight volumes covering the period up to June 1972. Literature cited includes urban runoff, storm water discharge, storm sewers, combined sewers, treatment methods, water reuse, legislation, and standards.

2. References

American Public Works Association. WATER POLLUTION ASPECTS OF URBAN RUNOFF: FINAL REPORT ON THE CAUSES AND REMEDIES OF WATER POLLUTION FROM SUFACE DRAINAGE. Washington, D.C.: U.S. Federal Water Pollution Control Administration, 1969. v, 272 p.

Study was conducted to determine the factors in the urban environment which contribute to the pollution potential of urban storm runoff water, and to develop methods which would limit this source of water pollution. Data included on the sources and amounts of storm water runoff pollution.

Burgess & Niple, Consulting Engineers. STREAM POLLUTION AND ABATEMENT FROM COMBINED SEWER OVERFLOWS. Water Pollution Control Research Series, DAST-32. Washington, D.C.: U.S. Federal Water Quality Administration, 1969. xiv, 197 p.

Report from a study on stream pollution resulting from combined sewer overflows, and the feasibility of alternate plans. Includes cost estimates. Various control methods are presented along with information on the degree of protection, and advantages and disadvantages.

Cincinnati. University of. Division of Water Resources. URBAN RUNOFF CHARACTERISTICS. Washington, D.C.: U.S. Environmental Protection Agency, Water Quality Office, 1970. Distributed by U.S. Government Printing Office. 341 p.

Provides detailed information on watershed characteristics and runoff quantity and quality from the Cincinnati area. Data was used to test and refine a storm water management model.

Field, Richard, et al. WATER POLLUTION AND ASSOCIATED EFFECTS FROM STREET SALTING. Environmental Protection Technology Series, R2-73-257. Washington, D.C.: U.S. Environmental Protection Agency, Office of Research and Monitoring, 1973. Distributed by U.S. Government Printing Office. ix, 48 p.

Report comprises a state-of-the-art review of highway deicing practices and associated environmental effects. Includes information on methods, equipment and materials used for snow and ice removal, on effects of deicing compounds on roadside soils, vegetation and trees, and on amounts of the compounds found in surface streams, public water supplies, and groundwater.

Grava, Sigurd. URBAN PLANNING ASPECTS OF WATER POLLUTION CONTROL. New York: Columbia University Press, 1969. xi, 223 p.

Analyzes the problem of waste disposal and the policy choices available to the urban planner. Discusses both the details of water pollution control and its planning implications, as well as its effect on community organization. Attention is given to the special problem of disposal in developing countries.

Kao, David T.Y., ed. PROCEEDINGS NATIONAL SYMPOSIUM ON URBAN RAINFALL AND RUNOFF AND SEDIMENT CONTROL. Lexington: Office of Research and Engineering Services, University of Kentucky, 1974. 246 p.

Most papers are technical descriptions of the process of runoff and of methods to alleviate this source of sediment. Legal and economic aspects are also included, as well as the application of models.

Leopold, Luna B. "The Hydrologic Effects of Urban Land Use." In MAN'S IMPACT ON ENVIRONMENT, edited by Thomas R. Detwyler, pp. 205-61. New York: McGraw-Hill, 1971.

Discusses the effects of urbanization on the hydrologic cycle which results because urban areas present large areas impervious to water. Urban areas are also a source of water pollution which reduced the amount of useful water.

McClanahan, Mark A., and Hoadley, Alfred W. A STUDY OF THE EFFECTS OF ISLAND DEVELOPMENT ON LAKE WATER QUALITY. Atlanta: Georgia Institute of Technology, Environmental Resources Center, and School of Engineering, 1972. vii, 49 p.

Results of a study which measured the chemical and physical parameters of a lake during the beginning phase of construction of a major recreational facility. The primary change occurring in the lake was an increase in siltation.

McGriff, E. Corbin, Jr. "The Effect of Urbanization on Water Quality." JOURNAL OF ENVIRONMENTAL QUALITY 1 (January-March 1972): 86-89.

Review of the effects of urbanization on water quality and hydrologic factors.

Pitt, Robert E., and Amy, Gary. TOXIC MATERIALS ANALYSIS OF STREET SURFACE CONTAMINANTS. Environmental Protection Technology Series, R2-73-283. Washington, D.C.: U.S. Environmental Protection Agency, Office of Research and Development, 1973. xi, 133 p.

Report of a study which examined the distribution of heavy metals on city streets. Study included analysis, size affinities, solubilities, and toxicities of road surface runoff mixture. Additional samples on rural roads and highways used for comparison.

Sartor, James D. WATER POLLUTION ASPECTS OF STREET SURFACE CONTAMINANTS. Environmental Protection Technology Series, R2-72-081. Washington, D.C.: U.S. Environmental Protection Agency, 1972. 240 p.

Study provides information which communities can utilize to evaluate the significance of street surface runoff as a water pollution source. Runoff was analyzed for various pollutants, including nutrients, heavy metals, and pesticides.

Terry, Robert C., Jr. ROAD SALT, DRINKING WATER, AND SAFETY: IMPROVING PUBLIC POLICY AND PRACTICE. Cambridge, Mass.: Ballinger Publishing, 1974. xxviii, 161 p.

Report to the Massachusetts Congress on the heavy utilization of deicing salt on roads and the subsequent contamination of drinking water sources. Presents the problem and recommendations for containing or controlling the problem.

U.S. Environmental Protection Agency. CHLORINATED MUNICIPAL WASTE TOXICITIES TO RAINBOW TROUT AND FATHEAD MINNOWS. Water Pollution Control Research Series, 18050 GZZ 10/71. Washington, D.C.: 1971. vii, 49 p.

Results of a study which assessed the effects of downstream chlorine residues on fish life and arrived at tolerance levels. Also determined the length of river below each plant outfall rendered unsuitable for resident fish populations.

3. Journal Articles

Adams, E.S. "Effects of Lead and Hydrocarbons from Snowmobile Exhaust on Brook Trout (Salvelinus Fontinalis). TRANSACTIONS AMERICAN FISHERY SOCIETY 104 (April 1975): 363-73.

Beszedits, S., and Netzer, A. "Program Will Control Pollution from Watercraft." WATER AND POLLUTION CONTROL 112 (June 1974): 33-34.

Caven, W.F.; Sirisinha, K.; and Lea, G. "Nitrogen Availability in Urban Runoff." JOURNAL WATER POLLUTION CONTROL FEDERATION 48 (February 1976): 339-45.

Cherkauer, D.S. "Urbanization Impact on Water Quality During a Flood in Small Watersheds." WATER RESOURCES BULLETIN 11 (October 1975): 987-98.

Dietrich, P., and Mulamoottil, G. "Does Recreational Use of Reservoirs Impair Water Quality?" WATER AND POLLUTION CONTROL 112 (February 1974): 16-18.

Doig, M.T. III, and Martin, D.F. "The Response of Gymnodinium Breve to Municipal Waste Materials." MARINE BIOLOGY 24 (March 1974): 223-28.

Field, R. "Coping with Urban Runoff in the United States." WATER RESEARCH 9 (May-June 1975): 499-505.

Field, R., and Lager, J.A. "Urban Runoff Pollution Control-State-of-the-Art." A.S.C.E. JOURNAL ENVIRONMENTAL ENGINEERING DIVISION 101 (February 1975): 107-25.

Oliver, B.G.; Milne, J.B.; and La Barre, N. "Chloride and Lead in Urban Snow." JOURNAL WATER POLLUTION CONTROL FEDERATION 46 (April 1974): 766-71.

Sartor, J[ames].D.; Boyd, G.B.; and Agardy, F.J. "Water Pollution Aspects of Street Surface Contaminants." JOURNAL WATER POLLUTION CONTROL FEDERATION 46 (March 1974): 458-67.

Scott, W.S. "The Effect of Road Deicing Salts on Sodium Concentration in an Urban Water-Course." ENVIRONMENTAL POLLUTION 10 (February 1976): 141-53.

Snoeyink, V.L., and Markus, F.I. "Chlorine Residues in Treated Effluents." WATER AND SEWAGE WORKS INCLUDING INDUSTRIAL WASTES 121 (April 1974): 35-38.

Wagenet, R.J., and Lawrence, C.H. "Recreational Effects on Bacteriological Quality of an Impounded Water Supply." JOURNAL OF ENVIRONMENTAL HEALTH 37 (July-August 1974): 16-20.

Whipple, W.; Hunter, J.; and Yu, S.L. "Unrecorded Pollution from Urban Runoff." JOURNAL WATER POLLUTION CONTROL FEDERATION 46 (May 1974): 873-85.

Chapter 17

TREATMENT AND MANAGEMENT OF WATER AND WASTEWATER—INCLUDING WATER QUALITY CRITERIA

The subject of water treatment, management, and reuse is too vast for exhaustive coverage within the scope of this book and the reader is referred to WASTE-WATER MANAGEMENT, p. 158, by Tchobanoglous, Smith, and Crites, for an in-depth treatment. This section focuses on general principles and sources of information on waste treatment for the nonengineer. The reader with minimal previous exposure to this topic is encouraged to review the citations provided in appendix B, "Selected Reading List," p. 239, under the topics of waste treatment and water renovation and reuse. The article by J.W.M. la Rivière, "A Critical View of Waste Treatment," p. 161, is highly recommended. The book by T.H.Y. Tebbutt, PRINCIPLES OF WATER QUALITY CONTROL, p. 158, is a very concise overview of water and water treatment for the nonengineer.

Journal articles included in this section were selected to provide the reader with several accessible current articles embracing the broad area of waste treatment, water renovation and reuse, and water quality criteria.

A. ABSTRACTS AND BIBLIOGRAPHIES

[U.S. Department of Interior.] U.S. Federal Water Pollution Control Administration. WATER POLLUTION CONTROL: WASTE TREATMENT AND WATER TREATMENT. Washington, D.C.: U.S. Government Printing Office, 1966. x, 126 p.

> Selected biological references on fresh and marine water related to topics in title.

[U.S. Department of Interior.] Water Resources Scientific Information Center. OZONE IN WATER AND WASTE WATER TREATMENT: A BIBLIOGRAPHY. Bibliography Series, WRSIC 74-204. Washington, D.C.: 1974. Distributed by National Technical Information Service. iv, 131 p.

> Contains references on the utilization of ozone for water purification and cleanup. References from SELECTED WATER RESOURCES ABSTRACTS up to 15 December 1973.

_____. A SELECTED ANNOTATED BIBLIOGRAPHY ON THE ANALYSIS OF WATER RESOURCES SYSTEMS. 6 vols. Vols. 1-2, Ithaca, N.Y.: Cornell University Water Resources and Marine Sciences Center, 1969, 1971; vols. 3-6, Washington, D.C.: U.S. Department of Interior, Water Resources Scientific Information Center, 1972-75.

Bibliographic citations pertinent to the applications of system analysis techniques for defining and evaluating alternative solutions to water resources problems.

_____. USE OF NATURALLY IMPAIRED WATER: A BIBLIOGRAPHY. Bibliography Series, WRSIC 73-217. Washington, D.C.: 1973. Distributed by National Technical Information Service. 364 p.

Bibliographic citations from SELECTED WATER RESOURCES ABSTRACTS through 15 February 1973. Most of the cited literature relates to the use of waters of high salinity in agriculture.

_____. WATER REUSE: A BIBLIOGRAPHY. 4 vols. Bibliography Series, WRSIC 73-215, 75-209. Washington, D.C.: 1973, 1975.

Bibliographic citations from SELECTED WATER RESOURCES ABSTRACTS through 15 January 1975, on the reuse of water by communities, industry and agriculture.

B. REFERENCES

American Water Works Association. WATER QUALITY AND TREATMENT: A HANDBOOK OF PUBLIC WATER SUPPLIES. 3d ed. New York: McGraw-Hill, 1971. xii, 654 p.

Text is concerned primarily with the methods involved in the purification and treatment of drinking water. In the discussion of the different methods, each method is evaluated as to usefulness, type, design, and operation.

Becker, Burton C., and Mills, Thomas R. GUIDELINES FOR EROSION AND SEDIMENT CONTROL PLANNING AND IMPLEMENTATION. EPA Technology Series, R2-72-015. Washington, D.C.: Environmental Protection Agency, Office of Research and Monitoring, 1972. Distributed by U.S. Government Printing Office. 228 p.

Prepared for those responsible for urban construction, this reference presents procedures for determining sediment and erosion problems, developing control plans, and for implementation. Information is also given on forty-two sediment and erosion control products (i.e., description of product and installation instructions).

Behrman, Abraham S. WATER IS EVERYBODY'S BUSINESS: THE CHEMISTRY OF WATER PURIFICATION. Garden City, N.Y.: Doubleday, 1968. viii, 229 p.

Nontechnical account of water supply and purification for drinking water.

Bennett, George W. MANAGEMENT OF LAKES AND PONDS. 2d ed. New York: Van Nostrand Reinhold, 1970. xx, 375 p.

Most of the book deals with the planning and management of aquatic habitats and how fish cohabit with the physical and biological characteristics of these habitats. Some information is provided on the effects of pesticides, phosphates, and other pollutants on aquatic life.

Benoit, Richard J. "Self-Purification in Natural Waters." In WATER AND WATER POLLUTION HANDBOOK, vol. 1, edited by Leonard L. Ciaccio, pp. 223-61. New York: Marcel Dekker, 1971.

Presentation of the physical, chemical, and biological processes involved in the self-purification of natural waters.

Boersma, Larry L.; Barlow, E.W.R.; and Rykbost, K.A. USE OF REACTOR COOLING WATER FROM NUCLEAR POWER PLANTS FOR IRRIGATION OF AGRICULTURAL CROPS. Water Resources Research Institute, Report, WRRI-12. Corvallis, Oreg.: Oregon State University, Water Resources Research Institute, 1972. xii, 95 p.

Report on the feasibility of utilizing waste heat for irrigation. Discusses ion uptake and growth response of wheat and corn seedlings.

Bolton, Ronald L., and Klein, Louis. SEWAGE TREATMENT: BASIC PRINCIPLES AND TRENDS. 2d rev. ed. Ann Arbor, Mich.: Ann Arbor Science Publishers, 1971. 256 p.

Presents techniques for the analysis and treatment of sewage and industrial wastes, especially as related to disposal and its relation to natural waters.

Boyle, Robert H.; Graves, John; and Watkins, T.H. THE WATER HUSTLERS. San Francisco: Sierra Club, 1971. 253 p.

Examination of the water plans of Texas, California, and New York and the attempts to control and transfer water.

Bradley, Charles C. "Human Water Needs and Water Use in America." SCIENCE 138 (26 October 1962): 489-91.

Indicates the water needs of man, the amount of water being used, and when we can expect to reach our limits.

Brahtz, J.F. Peel, ed. COASTAL ZONE MANAGEMENT: MULTIPLE USE WITH CONSERVATION. New York: John Wiley and Sons, 1972. xii, 352 p.

Written for the planner, this book discusses the conflicts and problems associated with various aspects of coastal zone development. Describes technologies available to achieve planning goals.

Briggs, Peter. WATER: THE VITAL ESSENCE. New York: Harper and Row, 1967. ix, 223 p.

Book relates popular stories related to water, i.e., the voyage of the CHALLENGER and fish kills on the Mississippi. Approximately half the book is devoted to the oceans with chapters relating to water pollution and to water management.

Camp, Thomas R., and Meserve, Robert L. WATER AND ITS IMPURITIES. 2d ed. New York: Reinhold Publishing, 1974. 384 p.

Physical and chemical properties of pure water and of impurities found in water are discussed. Text covers water standards, composition of natural and polluted water, microbial action in the breakdown of pollutants, and the dispersal properties of receiving waters. Technical.

Clark, B. David; Rittall, W.F.; Baumgartner, D.J.; and Bryan, K.V. BARGED OCEAN DISPOSAL OF WASTES: A REVIEW OF CURRENT PRACTICE AND METHODS OF EVALUATION. Corvallis, Oreg.: Environmental Protection Agency, Pacific Northwest Laboratory, 1971. ix, 120 p.

General but somewhat technical description of the types of wastes amenable to disposal in the ocean. Discussion includes the physical characteristics of various wastes plus transport and solution phenomena. Includes the economics of ocean disposal and the reported effects of past discharge operations.

Clark, John W.; Viessman, Warren, Jr.; and Hammer, Mark J. WATER SUPPLY AND POLLUTION CONTROL. 2d ed. International Series in Civil Engineering. Scranton, Pa.: International Textbook, 1971. xiii, 661 p.

Designed for a course in sanitary engineering and water resources, this text discusses development of water supplies, wastewater systems, treatment, water quality, various treatment processes, the reuse of wastewater, and legal considerations.

Cooper, Edwin L., ed. "Symposium on Water Quality Criteria to Protect Aquatic Life." In TRANSACTIONS OF THE AMERICAN FISHERIES SOCIETY, Special Publication no. 4. Washington, D.C.: American Fisheries Society, 1967. 37 p.

Symposium proceedings are primarily reviews on the effects of water pollution on aquatic life. Contains good reviews on the effects of dissolved oxygen and of thermal pollution on aquatic life.

Culp, Russell L., and Culp, Gordon L. ADVANCED WASTE-WATER TREATMENT. New York: Van Nostrand Reinhold, 1971. viii, 310 p.

Technical book which describes methods designed to remove pollutants normally remaining after conventional wastewater treatment. Indicates those processes which are already in use including their reliability, performance and cost, and those processes which probably will be used in the future.

Dawson, G.W.; Shuckrow, A.J.; and Swift, W.H. CONTROL OF SPILLAGE OF HAZARDOUS POLLUTING SUBSTANCES. Water Pollution Control Research Series. Washington, D.C.: U.S. Federal Water Quality Administration, 1970. xvii, 350 p.

Report on over eight hundred chemicals which were evaluated as to their production and transport quantities, their critical concentration which result in water quality impairment, and their control and removal methods. Contents also include legal and enforcement considerations and contingency planning.

Doudoroff, Peter, and Shumway, Dean L. DISSOLVED OXYGEN REQUIREMENTS OF FRESHWATER FISHES. FAO Fisheries Technical Paper no. 86. Rome: Food and Agriculture Organization of the United Nations, 1970. xi, 291 p.

Critical review and interpretation of the data related to the effects of varying levels of dissolved oxygen on fish. Includes lethal levels and effect on fecundity, embryonic development, growth, swimming, metabolism, and behavior.

Dougal, Merwin D.; Baumann, E. Robert; and Timmons, John F. PHYSICAL AND ECONOMIC FACTORS ASSOCIATED WITH THE ESTABLISHMENT OF STREAM WATER QUALITY STANDARDS. 3 vols. Iowa State Water Resources Research Institute, Project no. A-001-1A. Ames: Sanitary Engineering Section, Engineering Research Institute, Iowa State University, 1970.

Results of a comprehensive study of water pollution control and stream water quality as they relate to the establishment of stream and effluent standards. Volume 1 is a review of the findings, while volumes 2 and 3 present the data for a water quality model. Data was collected on Iowa streams.

Dunseth, Maria G.; Salutsky, M.C.; Ries, K.M.; and Shapiro, J.J. ULTIMATE DISPOSAL OF PHOSPHATE FROM WASTE WATER BY RECOVERY AS FERTILIZER. Water Pollution Control Research Series. Washington, D.C.: U.S. Government Printing Office, 1970. 74 p.

Technical description of the treatment of sewage to remove phosphate for use as a fertilizer.

Eckenfelder, W. Wesley, Jr. WATER QUALITY ENGINEERING FOR PRACTICING ENGINEERS. New York: Barnes and Noble, 1970. xii, 328 p.

Overview of water pollution, i.e., characterization of pollutants, analysis, types of pollutants from municipalities, and industry. Includes a discussion of the various treatments and costs associated with each treatment.

Ehlers, Victor M., and Steel, Ernest W. MUNICIPAL AND RURAL SANITATION. 6th ed. McGraw-Hill Series in Sanitary Science and Water Resources Engineering. New York: McGraw-Hill, 1965. xiv, 663 p. Glossary.

Primarily concerned with the public health aspects of water treatment. Includes information on the sanitation of refuse, food products, and the control of insect and rodent vectors. Data on maximum permissible radioisotope concentrations in water is included.

Furon, Raymond. THE PROBLEM OF WATER: A WORLD STUDY. Translated by Paul Barnes. New York: American Elsevier Publishing, 1967. xvi, 208 p.

Examines the general aspects of water such as the importance of water, saltwater, freshwater, and the use of water by agriculture, industry, and people. Particular attention is given to water use, past and present. Sections are also included on where water is located and on the development of new water supplies.

Gloyna, Earnest F., and Butcher, William S., eds. CONFLICTS IN WATER RESOURCES PLANNING. Water Resources Symposium, no. 5. Austin: University of Texas, Center for Research in Water Resources, 1972. x, 172 p.

Collection of papers presented at a symposium on water management. Topics include the government's role in resources planning, the relation of water resources planning to land use policy, and the use of ecological principles as a tool in planning.

Gloyna, Earnest F., and Eckenfelder, W. Wesley, Jr., eds. ADVANCES IN WATER QUALITY IMPROVEMENT. Austin: University of Texas Press, 1968. xviii, 513 p.

Series of papers presented at a symposium encompassing various aspects of water quality. One section covers stream pollution from an economic, ecological, and physical perspective, while two other sections are concerned with waste treatment, primarily new concepts in biological waste treatment, and current trends in low cost treatment.

Gunnerson, Charles G. AN APPRAISAL OF MARINE DISPOSAL OF SOLID WASTES OFF THE WEST COAST: A PRELIMINARY REVIEW AND RESULTS OF A SURVEY. Washington, D.C.: U.S. Bureau of Solid Waste Management, 1970. 32 p. Pamphlet.

Discussion of preliminary results relating to the disposal of solid wastes in marine waters.

Hawkes, Herbert A. THE ECOLOGY OF WASTE WATER TREATMENT. New York: Macmillan, 1961. ix, 203 p.

Presents information for the nonbiologist on the implementation and maintenance of biological waste treatment, i.e., activated sludge. Contents include, the ecology of activated sludge, application of ecological principles to wastewater treatment, and ecological considerations in the design and operation of activated sludge plants.

Hopkins, Edward S.; Bingley, W. McLean; and Schucker, George W. THE PRACTICE OF SANITATION IN ITS RELATION TO THE ENVIRONMENT. 4th ed. Baltimore: Williams and Wilkins, 1970. xi, 550 p.

Encompasses the sanitation of water, the disposal of sewage, refuse and solid wastes, and insect and rodent control. A chapter on controlling stream pollution gives a general overview including the effects of pollutants on aquatic life, eutrophication, economic factors, and critical levels of some pollutants.

International Conference on Water Pollution Research. ADVANCES IN WATER POLLUTION RESEARCH. 6 vols. New York: Pergamon Press, 1962-72.

Proceedings of conferences held every other year. Each volume has its individual editor. The last two conferences (i.e., 1970, 1972) have included section titles on such topics as freshwater problems, thermal pollution, trace metals, and modeling. Conferences cover wide variety of topics from effect of pollutants on aquatic organisms to various aspects of waste treatment.

Klein, Louis. RIVER POLLUTION. Vol. 3: CONTROL. London: Butterworth, 1966. xv, 484 p.

Treatment of wastewaters, detection and measurement of pollutants, and the pollution of tidal and coastal waters are some of the topics covered.

Kneese, Allen V., and Bower, Blair T. MANAGING WATER QUALITY: ECONOMICS, TECHNOLOGY, INSTITUTIONS. Resources for the Future. Baltimore: Johns Hopkins Press, 1968. x, 328 p.

Contents include the nature of waste discharges, their effects on receiving waters and the management of waste loads. Also includes a discussion of resource allocation, cost involved in management and policy alternatives. Text emphasizes water management and the economics of water policies.

Kneese, Allen V., and Smith, Stephen C., eds. WATER RESEARCH. Baltimore: Johns Hopkins Press, 1966. vii, 526 p.

Papers presented at the Seventh Water Resources Conference at Colorado State University, covering the following areas: economic aspects of water management and cleanup, the politics and law of water development, and hydrology and engineering.

League of Women Voters. THE BIG WATER FIGHT: TRIALS AND TRIUMPHS IN CITIZEN ACTION ON PROBLEMS OF SUPPLY, POLLUTION, FLOODS, AND PLANNING ACROSS THE USA. Brattleboro, Vt.: Stephen Green Press, 1966. ix, 246 p. Glossary, reading list.

General treatment of water supply problems, pollution, floods, and planning from examples taken from various part of country.

McKee, Jack E., and Wolf, Harold W., eds. WATER QUALITY CRITERIA. 2d ed. Sacramento: California State Water Resources Control Board, 1963. 548 p. Bibliography.

First section of this book presents a general working definition of pollution, as well as water quality criteria developed by various states and judicial rulings affecting these criteria. A major portion of the book is a listing of potential and actual pollutants with information on their source and on toxicity. Good source of information up to 1963.

Moss, Frank E. THE WATER CRISES. New York: F.A. Praeger, 1967. xiii, 305 p.

Written from a political viewpoint, this book discusses the history and politics of water use and water pollution. Possible political and social solutions are given.

National Academy of Sciences. WASTES MANAGEMENT CONCEPTS FOR THE COASTAL ZONE; REQUIREMENTS FOR RESEARCH AND INVESTIGATION. Report 1855. Washington, D.C.: 1970. vii, 126 p. Bibliography.

Indicates, in general, what is known and what needs to be done to establish an effective management program of waste discharge into coastal areas. Topics discussed are the physical processes in receiving waters, chemical factors and biological effects of waste discharge.

Nikolaieff, George A., ed. THE WATER CRISIS. Reference Shelf. New York: H.W. Wilson, 1967. 192 p. Bibliography.

Contains reprints of articles, excerpts from books and addresses which address the problem of water use and shortage, and ways of obtaining more water.

O'Laoghaire, D.T., and Himmelbau, D.M. OPTIMAL EXPANSION OF A WATER RESOURCES SYSTEM. Water Pollution: A Series of Monographs, vol. 1. New York: Academic Press, 1974. xiii, 273 p.

Describes methodology for water resources planning, including optimization strategy and sensitivity analysis. Intended as an aid to policymakers. Technical.

Parizek, R.R., et al. WASTEWATER RENOVATION AND CONSERVATION. Pennsylvania State Studies, no. 23. University Park: Pennsylvania State University, 1967. 71 p.

Report of the problems, methods, results, and conclusions of a study which sprayed sewage plant effluent on croplands and forested areas in an attempt to establish a cycle of renovation.

Phelps, Earle B. STREAM SANITATION. New York: John Wiley and Sons, 1944. xi, 276 p.

Although somewhat dated, this book is a classic. It discusses some of the principal interests that have developed in streams from the single viewpoint of sanitation. Deals with a wide number of topics.

Platt, Rutherford H. WATER: THE WONDER OF LIFE. Englewood Cliffs, N.J.: Prentice-Hall, 1971. xi, 274 p.

Discussion of the importance of water in the origin, development and balance of life, as well as the threat of water quality degradation.

Rosen, S.J. MANUAL FOR ENVIRONMENTAL IMPACT EVALUATION. Englewood Cliffs, N.J.: Prentice-Hall, 1976. 240 p.

Manual provides an understanding of the National Environmental Policy Act, its philosophy and requirements. Author provides practical suggestions on techniques, procedures, and methodology to those preparing, reviewing, or responsible for purchasing environmental assessments.

Rubin, Alan J., ed. CHEMISTRY OF WATER SUPPLY, TREATMENT AND DISTRIBUTION. Ann Arbor, Mich.: Ann Arbor Science Publishers, 1974. viii, 446 p.

Compilation of papers, presented at a symposium, dealing primarily with water analysis and the removal of compounds from water. Several papers on chlorine and its effect in water are also included. Technical.

Swenson, H.A., and Baldwin, H.L. A PRIMER ON WATER QUALITY. Washington, D.C.: U.S. Government Printing Office, 1965. iii, 27 p.

Brief, nontechnical overview on water quality, covering the characteristics of water, its treatment and criteria for safe water.

Sykes, G., and Skinner, F.A., eds. MICROBIAL ASPECTS OF POLLUTION. New York: Academic Press, 1971. xiii, 289 p.

Sixteen contributions include consideration of the health hazards of pollution, the consequences of sewage pollution of water supplies and the safe disposal of infective materials. Series of papers deal with the problem of water purification and disposal of sewage, industrial wastes, and the wastes of new industrial products and their effect on rivers and lakes.

Symons, James M., ed. WATER QUALITY BEHAVIOR IN RESERVOIRS: A COMPILATION OF PUBLISHED RESEARCH PAPERS. Public Health Service Publication no. 1930. Cincinnati: U.S. Department of Health, Education and Welfare, 1969. xii, 616 p.

Covers the influence of impoundment on water quality with emphasis on the problems caused by stratification. A number of papers discuss methods for destratification and their effect on aquatic biota. Other papers are concerned with dissolved oxygen levels and the nitrogen cycle in reservoirs.

Tchobanoglous, George; Smith, Robert; and Crites, Ronald, eds. WASTEWATER MANAGEMENT: A GUIDE TO INFORMATION SOURCES. Man and the Environment Series, vol. 2. Detroit: Gale Research Co., 1976. xiii, 202 p.

An authoritative and comprehensive work on wastewater management that brings together a diverse set of information sources from the physical, social and natural sciences.

Tebbutt, T.H.Y. PRINCIPLES OF WATER QUALITY CONTROL. New York: Pergamon Press, 1971. ix, 179 p.

Very concise overview of water, detailing the characteristics of polluted and nonpolluted waters, and the methods involved in water and waste water treatment such as sedimentation, coagulation, aerobic and anaerobic, oxidation, disinfection, and so forth.

United Nations World Health Organization. INTERNATIONAL STANDARDS FOR DRINKING WATER. 3d ed. Geneva: 1971. 70 p.

Presents international standards and minimum requirements for the bacterial and chemical quality of domestic water supplies. Includes methods of analysis to determine water quality.

U.S. Council on Environmental Quality. NATIONAL OIL AND HAZARDOUS MATERIALS POLLUTION CONTINGENCY PLAN. Washington, D.C.: 1970. iii, 15 p.

Proposal for the efficient coordination and effective action to minimize damage from oil and other hazardous materials.

[U.S. Department of Interior.] U.S. Federal Water Pollution Control Administration. WATER QUALITY CRITERIA. Washington, D.C.: 1968. x, 234 p.

Contains recommendations of water quality criteria for recreation and aesthetic uses as well as industrial and agricultural uses. Extremely useful book for aquatic biologists, environmental specialists, water quality specialists, those preparing environmental impact statements, and engineers.

U.S. Environmental Protection Agency. CONTROL OF HAZARDOUS MATERIAL SPILLS: PROCEEDINGS. Environmental Sourcebook Series, vol. 1. Washington, D.C.: Graphic Management Corporation, 1972. 223 p.

Compilation of papers presented at a conference, most dealing with prevention and containment of hazardous materials spills. Those dealing with water, cover treatment after spills and the effect of spills on aquatic organisms.

______. CONTROL OF OIL AND OTHER HAZARDOUS MATERIALS. Washington, D.C.: 1972. Various pagination.

Guide to various aspects of oil spill problems includes coverage of the effects of spills, their sources, control and legislation.

U.S. National Water Commission. WATER POLICIES FOR THE FUTURE: FINAL REPORT TO THE PRESIDENT AND TO THE CONGRESS OF THE UNITED STATES. Washington, D.C.: U.S. Government Printing Office, 1973. xxvii, 579 p. Glossary.

Report examines the problems of reconciling federal and state water policies, of integrating ground and surface water management, and the manner in which water can be more efficiently used.

______. WATER POLLUTION CONTROL IN THE UNITED STATES. Arlington, Va.: 1971. Distributed by National Technical Information Service. 259 p.

Discussion of the various aspects of water pollution including responsibility and public impact, water quality goals, objectives of water management, water standards, costs, and effectiveness of present programs. Panel also gave recommendations for the implementation of an effective water quality control program.

U.S. Office of Water Quality. GUIDELINES, WATER QUALITY MANAGEMENT PLANNING. Washington, D.C.: 1971. Various pagination.

Presents the preliminary guidelines established by the Environmental Protection Agency for management of water quality. Includes metropolitan and regional plans, approaches to management, and evaluation procedures.

Velz, Clarence J. APPLIED STREAM SANITATION. Environmental Science and Technology Series. New York: Wiley-Interscience, 1970. ix, 619 p.

Text covers the natural ability of streams to purify themselves and the impact of waste disposal and other processes on this ability. Methods are presented on the measurement of the cleansing ability of streams and how to efficiently utilize this ability. There are appendixes on statistical analysis and bacterial enumeration.

White, Gilbert F. STRATEGIES OF AMERICAN WATER MANAGEMENT. Ann Arbor: University of Michigan Press, 1969. xii, 155 p.

Examines water resources in the United States and the strategies which we have used, and are using, to manage water supplies. Indicates who makes the decision and the results.

Wright, James C. THE COMING WATER FAMINE. New York: Coward-McCann, 1966. 255 p.

Popular account of the water shortage and of water pollution which reduces the amount of usable water.

Zajic, James E. WATER POLLUTION: DISPOSAL AND REUSE. 2 vols. New York: Marcel Dekker, 1971.

Analyzes the problem of water pollution from a treatment aspect in this reference. Includes a discussion of the biological parameters of water quality, the characteristics of streams, the control of aquatic plants, and eutrophication. Primary emphasis is on the various biological, chemical, and physical aspects of water purification.

C. JOURNAL ARTICLES

Allen, J. "Sewage Farming." ENVIRONMENT 15 (March 1973): 36-41.

DeRouen, T.A., and Diem, J.E. "The New Orleans Drinking Water Controversy: A Statistical Perspective." AMERICAN JOURNAL OF PUBLIC HEALTH 65 (October 1975): 1060-62.

Dworkin, D.M. "Water Reuse: A Flexible and Efficient Management Alternative for Municipal Water Supply." WATER RESOURCES BULLETIN 11 (October 1975): 607-15.

Eckenfelder, W.W[esley]., Jr. "Wastewater Treatment Design: Economics and Techniques: Part I." WATER AND SEWAGE WORKS 122 (June 1975): 62-65.

English, J.N., and Mitchell, T.M. "Water Reclamation and Reuse." JOURNAL WATER POLLUTION CONTROL FEDERATION 48 (June 1976): 1174-80.

Gilbert, R.G., et al. "Wastewater Renovation and Reuse: Virus Removal by Soil Filtration." SCIENCE 192 (June 1976): 1004-50.

Kohlhaas, C.A. "Focus--Comprehensive Water Resources Management." WATER AND SEWAGE WORKS 122 (June 1975): 70-75.

la Rivière, J.W.M. "A Critical View of Waste Treatment." In WATER POLLUTION MICROBIOLOGY, edited by Ralph Mitchell, pp. 365-88. New York: Wiley-Interscience, 1972.

Marx, J.L. "Drinking Water: Another Source of Carcinogens?" SCIENCE 186 (November 1974): 809-11.

Oleckno, W.A. "Predicting the Water Pollution Potential of Proposed Sanitary Landfills. Part I: Sanitary Landfill Leachate--What It Is. Part II: An Index of the Water Pollution Potential of Sanitary Landfills." JOURNAL ENVIRONMENTAL HEALTH 38 (March-April 1976): 331-33, 334-36.

Osgood, J.O. "Ground-Water's Role in Water Quality Management." A.S.C.E. JOURNAL HYDRAULICS DIVISION 101 (March 1975): 517-21.

Page, T.; Harris, R.H.; and Epstein, S.S. "Drinking Water and Cancer Mortality in Louisiana." SCIENCE 193 (July 1976): 55-57.

Pickering, R.J., and Ficke, J.F. "Design of Nationwide Water--Quality-Monitoring Networks." JOURNAL AMERICAN WATER WORKS ASSOCIATION 68 (February 1976): 82-85.

Reinhardt, A.W.; Spath, D.P.; and Jopling, W.F. "Organics, Water and Health: A Reuse Problem." JOURNAL AMERICAN WATER WORKS ASSOCIATION 67 (September 1975): 477-80.

Schmidt, C.J.; Kugelman, I.; and Clements, E.V. III. "Municipal Wastewater Reuse in the U.S." JOURNAL WATER POLLUTION CONTROL FEDERATION 47 (September 1975): 2229-45.

Schwinn, D.E., and Tozer, G.K. "Largest Advanced Wastewater Treatment Plant in the U.S. and in the World." ENVIRONMENTAL SCIENCE AND TECHNOLOGY 8 (October 1974): 892-97.

Sylvester, J.R. "Biological Considerations on the Use of Thermal Effluents for Finfish Aquaculture." AQUACULTURE 6 (July 1975): 1-10.

Tilley, L.J., and Haushild, W.L. "Use of Productivity of Periphyton to Estimate Water Quality." JOURNAL WATER POLLUTION CONTROL FEDERATION 47 (August 1975): 2157-71.

Truett, J.B.; Johnson, A.C.; Rowe, W.D.; Feigner, K.D.; and Manning, L.J. "Development of Water Quality Management Indices." WATER RESOURCES BULLETIN 11 (June 1975): 436-45.

Whipple, W., Jr. "Water Quality Planning for the Delaware Estuary--An Example." WATER RESOURCES BULLETIN 11 (April 1975): 300-305.

Chapter 18

SOCIAL, ECONOMIC, AND POLITICAL ASPECTS OF WATER POLLUTION

Ultimate control of water pollution is dependent not on our knowledge of the pollutant, its source, toxicity or treatment, but on the complex interaction of social attitudes, economic benefits, and administrative implementation of laws and regulations. Until recently, the disposal of wastes into rivers, lakes, and the ocean was standard procedure, dilution of toxic material being the accepted mode of treatment. Since the 1960s there has been growing social and political pressure for cleaner water, spurred in part by the increasing and competing uses for water. Some of the initial enthusiasm, however, has diminished due to the economic reality of implementing clean water standards.

The following references provide an introduction to the economics of water treatment, and to the laws and administrative structure needed for implementation. The reference by Dorfman and Jacoby, p. 166, is a good presentation of the economic and political aspects involved in water pollution control.

Other volumes in the Man and the Environment Information Guide Series (ENVIRONMENTAL ECONOMICS, Field and Willis, 1979; ENVIRONMENTAL LAW, Schwartz, 1977; ENVIRONMENTAL VALUES, 1960-1972, Owings, 1976) offer more in-depth coverage of this area.

A. ABSTRACTS AND BIBLIOGRAPHIES

Field, Barry C., and Willis, Cleve E., eds. ENVIRONMENTAL ECONOMICS: A GUIDE TO INFORMATION SOURCES. Man and the Environment Series, vol. 8. Detroit: Gale Research Co., 1979. xix, 243 p.

> Annotated bibliography of both theoretical and empirical studies. Includes chapter on economic studies relating to water quality.

Hamilton, Henry R.; Owens, D.H.; Carroll, T.E.; Glenn, A.R.; and Gilmour, B.A. BIBLIOGRAPHY ON SOCIO-ECONOMIC ASPECTS OF WATER RESOURCES. Washington, D.C.: U.S. Department of Interior, Office of Water Resources Research, 1966. xii, 453 p.

Focuses on the period from 1955 through 1965. Includes bibliographic citations with short abstracts. Areas covered include: supply and demand for water, methods and applications of cost benefit analysis, methods of determining economic values, economic impact of water resources, and values of sportfisheries and wildlife.

Jacobstein, J. Myron, and Mersky, Roy M. WATER LAW BIBLIOGRAPHY 1847-1965. Silver Spring, Md.: Jefferson Law Book, 1966. xvi, 249 p. Supplement 1: 1966-67, 1969. 138 p.

Concerned primarily with water utilization. Includes a section on legal aspects of water pollution.

Morrison, Denton E.; Hornback, Kenneth E.; and Warner, W. Keith, eds. ENVIRONMENT: A BIBLIOGRAPHY OF SOCIAL SCIENCE AND RELATED LITERATURE. Washington, D.C.: U.S. Environmental Protection Agency, Office of Research and Monitoring, 1974. Distributed by U.S. Government Printing Office, vii, 860 p.

Bibliography of nearly five thousand items for the period up to and including 1974, covering the literature in the fields of communications, economics, education, and history. Includes sections on water management, use, pollution, and politics.

Owings, Loren C., ed. ENVIRONMENTAL VALUES, 1860-1972: A GUIDE TO INFORMATION SOURCES. Man and the Environment Series, vol. 4. Detroit: Gale Research Co., 1976. xii, 324 p.

Annotated bibliography of materials which trace the development of attitudes toward nature in the United States, including travel reports, literature on landscape painting, nature writing, nature study, conservation, camping and the "back to nature" movement.

Schwartz, Mortimer D., ed. ENVIRONMENTAL LAW: A GUIDE TO INFORMATION SOURCES. Man and the Environment Series, vol. 6. Detroit: Gale Research Co., 1977. xiii, 191 p.

Annotations of books, government documents and published conference proceedings or papers dealing with environmental law as pertinent to the protection, preservation and rehabilitation of the physical environment. Includes chapters on water pollution control, pesticide regulation, fisheries conservation and conservation of marine resources.

Tihansky, Dennis P. COST ANALYSIS OF WATER POLLUTION CONTROL: AN ANNOTATED BIBLIOGRAPHY. Socioeconomic Environmental Studies Series, EPA R5-73-017. Washington, D.C.: U.S. Environmental Protection Agency, Office of Research and Monitoring, 1973. Distributed by U.S. Government Printing Office. v, 393 p.

Bibliography of nearly two thousand publications, unpublished papers, and theses from between 1960 and 1972, which cover the economics of water pollution control.

U.S. Advisory Commission on Intergovernmental Relations. THE QUEST FOR ENVIRONMENTAL QUALITY: FEDERAL AND STATE ACTION 1969-1970, ANNOTATED BIBLIOGRAPHY. Washington, D.C.: U.S. Government Printing Office, 1971. v, 63 p.

Report on federal and state action relating to environmental problems along with an annotated bibliography covering government actions and the economic, political, and social implications of pollution and its control.

B. REFERENCES

Clements, Walter C.; Olson, Roy H.; and Simpson, Bruce L., eds. THE COMMON LAW OF WATER. LIMNOS 4 (Fall 1971): 1-36

Discussion of basic water rights and their relation to water pollution. Calls attention to important federal legal decisions and to those of Illinois.

Conference on Law, Science and Politics: Water Pollution and Its Effects Considered as a World Problem. WATER POLLUTION AS A WORLD PROBLEM: THE LEGAL, SCIENTIFIC AND POLITICAL ASPECTS. London: Europa for the David Davies Memorial Institute of International Studies, 1971. ix, 240 p.

Discussions of nuclear and thermal pollution, oil pollution, chemical, and pesticide pollution, and on industrial pollution by a scientist, an administrator, and a lawyer. The treatment of each topic is general and reflects the interest of the speaker, i.e., legal, administrative, or scientific.

Cooley, Richard A., and Wandesforde-Smith, Geoffrey, eds. CONGRESS AND THE ENVIRONMENT. Seattle: University of Washington Press, 1970. xix, 277 p.

Series of papers which examine various environmental problems, including water pollution and the attempts to solve them by legislation. Papers relating to water pollution are: water projects and recreational benefits, water quality, wild and scenic rivers, and international control of the ocean floor.

Crocker, Thomas D., and Rogers, A.J. III. ENVIRONMENTAL ECONOMICS. Hinsdale, Ill.: Dryden Press, 1971. vii, 150 p.

General and popular treatment of the economic aspects of pollution, emphasizing the economic structure of the United States and its relationship to pollution. Good introduction to environmental economics for the noneconomist.

Davies, J. Clarence III, and Davies, Barbara S. THE POLITICS OF POLLUTION. 2d ed. Indianapolis: Pegasus Press, 1975. x, 254 p.

Book describes government policymaking procedures and analyzes the interests and ideas competing for dominance over pollution control. Discusses each major air and water pollution act passed by Congress and the controversies surrounding these decisions are examined. Also examines the forces which shape pollution policy.

Davis, Robert K. THE RANGE OF CHOICE IN WATER MANAGEMENT: A STUDY OF DISSOLVED OXYGEN IN THE POTOMAC ESTUARY. Resources for the Future. Baltimore: Johns Hopkins Press, 1968. 196 p.

Analysis of the processes involved in water quality management, both economic as well as administrative, as they pertained to the specific case of the Potomac Estuary. Useful appendixes on modeling.

Deran, Elizabeth Y. POLLUTION CONTROL: PERSPECTIVES ON THE GOVERNMENT ROLE. New York: Tax Foundation, 1971. 46 p.

Deals with all types of pollution, in general terms, and the possible means of inducing industry and municipalities to clean up pollution. Discusses the pros and cons of various fiscal proposals as well as technical, political, and economic difficulties faced in setting optimum control standards. Study brings together a number of cost estimates to provide a general view of the magnitude of the dollar amounts involved in making specified improvements.

Dorfman, Robert, and Jacoby, H. "A Public Decision Model Applied to a Local Pollution Problem." In ECONOMICS OF THE ENVIRONMENT, edited by Robert Dorfman and Nancy S. Dorfman, pp. 205-49. New York: Norton, 1972.

Describes a hypothetical river on which is located an industry and two towns, and the economic and political choices involved in cleaning it up. While the described situation is simple, it points up many of the controversies and complexities involved in "real-world" decision making. Good article.

Dow Chemical Company. AN ECONOMIC ANALYSIS OF EROSION AND SEDIMENT CONTROL METHODS FOR WATERSHEDS UNDERGOING URBANIZATION. Midland, Mich.: 1972. viii, 181 p.

Report of a study which computed the economic benefits of erosion control versus expenditures to control erosion in an urban area. Costs for various types of control and the amount of sediment controlled were estimated as was the cost of uncontrolled erosion.

Ducsik, Dennis W., ed. POWER, POLLUTION, AND PUBLIC POLICY. Massachusetts Institute of Technology Report, no. 24. Cambridge: MIT Press, 1971. xiii, 322 p.

Discussion of the environmental problems associated with electric power production and its affect on shoreline recreation, including possible solutions, both technological and political. Primarily concerned with the New England and Boston area.

Field, Donald R.; Barron, James C.; and Long, Burl F., eds. WATER AND COMMUNITY DEVELOPMENT: SOCIAL AND ECONOMIC PERSPECTIVES. Ann Arbor, Mich.: Ann Arbor Science Publishers, 1974. viii, 302 p.

Discussion of water utilization by communities as it relates to community development with both social and economic aspects considered. Impact of water use for recreation is also examined.

Fontana, Joseph, and Wulf, Norman A., eds. POLLUTION AND THE LAW. Washington, D.C.: Council on Natural Resources and Lands of the Federal Bar Association, 1971. v, 28 p.

Focuses on federal activities in the area of pollution control. Designed for the high school student.

Freeman, A. Myrick III, and Haveman, Robert H. "Water Pollution Control, River Basin Authorities, and Economic Incentives: Some Current Policy Issues." PUBLIC POLICY 19 (Winter 1971): 53-74.

Discussion of the possibilities for controlling water pollution through economic incentives and for administering water systems on a basin approach.

Goldman, Marshall I., ed. CONTROLLING POLLUTION: THE ECONOMICS OF A CLEANER AMERICA. Englewood Cliffs, N.J.: Prentice-Hall, 1967. xiii, 175 p.

Collection of general articles which covers the following three areas of the pollution problem: "the nature of the problem," "economic analysis," and "current cases." Many articles are from SATURDAY REVIEW and involve both air and water pollution.

______. ECOLOGY AND ECONOMICS: CONTROLLING POLLUTION IN THE 70'S. Englewood Cliffs, N.J.: Prentice-Hall, 1972. xi, 234 p.

Series of articles discussing the problems associated with pollution, primarily economic and political, attempts at solving these problems, and thoughts about future goals.

Grad, Frank P.; Rathjens, George W.; and Rosenthal, Albert J. ENVIRONMENTAL CONTROL: PRIORITIES, POLICIES AND THE LAW. New York: Columbia University Press, 1971. viii, 311 p.

Examination of the issues involved in a federal environmental policy, what the priorities and goals should be, how these goals can be accommodated within the federal-state structure, and whether the government can enforce its goals.

Graham, Frank, Jr. DISASTER BY DEFAULT: POLITICS AND WATER POLLUTION. New York: M. Evans & Co. 1966. Distributed in association with Lippincott. 256 p.

Compilation of case histories involving water pollution, detailing the politics, administration, and costs involved.

Hite, Jame C.; Macaulay, H.H.; Stepp, J.M.; and Yandle, B., Jr. THE ECONOMICS OF ENVIRONMENTAL QUALITY. Domestic Affairs Studies, no. 5. Washington, D.C.: American Enterprise Institute for Public Research, Research, 1972. 113 p.

Good discussion of the economics of pollution abatement provides background information about the environment and about economics. Presents the problems associated with environmental quality, economic demand, and the allocation of scarce resources. Good introduction for the noneconomist.

Hurley, William D. ENVIRONMENTAL LEGISLATION. Springfield, Ill.: Charles C Thomas, 1971. ix, 81 p.

Provides a detailed description of the major laws enacted by the U.S. Congress to cope with pollution and analyzes their effect. Emphasis is on legislation and how well it works in curbing pollution.

Kalter, Robert J. THE ECONOMICS OF WATER-BASED OUTDOOR RECREATION: A SURVEY AND CRITIQUE OF RECENT DEVELOPMENTS. U.S. Army Engineers Institute for Water Resources Report, no. 71-8. Washington, D.C.: U.S. Army Engineer Institute, 1971. 192 p.

Report of a study which surveyed and critiqued the literature and investigated the practice of estimating use and benefits from outdoor recreation. Methods are outlined and statistical problems delineated for determining use and benefits.

Kneese, Allen V. APPROACHES TO REGIONAL WATER QUALITY MANAGEMENT. Washington, D.C.: Resources for the Future, 1967. 47 p.

Presents information on legal and institutional approaches to water quality management in various parts of the world and assesses these regional programs, primarily in terms of economic efficiency.

______. WATER POLLUTION: ECONOMIC ASPECTS AND RESEARCH NEEDS. Washington, D.C.: Resources for the Future, 1962. 107 p.

Water pollution is examined from an economic framework and areas are outlined where research is needed to better define the effects and results of water pollution. Examines scientific, technological, and economic aspects of controlling water pollution.

Kneese, Allen V.; Ayres, Robert U.; and d'Arge, Ralph C. ECONOMICS AND THE ENVIRONMENT. Resources for the Future. Baltimore: Johns Hopkins Press, 1970. x, 120 p.

Pollution is analyzed from the perspective of developing an economic model which could be used to derive environmental taxes for pollution control. Technical.

Kneese, Allen V., and Schultze, Charles L. POLLUTION, PRICES, AND PUBLIC POLICY. Washington, D.C.: Brookings Institute, 1975. 125 p.

Summarizes the economic and technical background necessary to understand the pollution control problem. The legislative history of federal pollution control efforts is reviewed, as well as the costs and enforcement difficulties. Alternative policies are proposed.

Kneese, Allen V.; Rolfe, Sidney E.; and Harned, J.W., eds. MANAGING THE ENVIRONMENT: INTERNATIONAL ECONOMIC COOPERATION FOR POLLUTION CONTROL. New York: Praeger Publishers, 1971. xxxvii, 356 p.

Outgrowth of a conference on the economics of pollution in the United States and the world. Also presents the technological state-of-the-art. Contains two studies on the costs of cleaning up water systems plus associated legal aspects (i.e., federal laws and court rulings).

McCaull, Julian. "Who Owns the Water?" ENVIRONMENT 12 (October 1970): 30-39.

Surveys the control of water pollution on an international and regional level with emphasis on the problems arising when rivers cross political boundaries. Some attempted solutions are presented.

Murphy, Earl F. WATER PURITY: A STUDY IN LEGAL CONTROL OF NATURAL RESOURCES. Madison: University of Wisconsin Press, 1961. x, 212 p.

Discussion of the history of laws relating to the control of water resources. Follows the development of administrative control of water quality in Wisconsin.

National Association of Counties. Research Foundation. COMMUNITY ACTION PROGRAM FOR WATER POLLUTION CONTROL. Washington, D.C.: 1966. 182 p. Bibliography.

A "how to clean up your water" book designed for community leaders. Contents include: enabling legislation for pollution control programs, plus planning, organizing and financing control programs, and construction and staffing of facilities and public relations.

Scott, David L. POLLUTION IN THE ELECTRIC POWER INDUSTRY: ITS CONTROL AND COSTS. Lexington, Mass.: Lexington Books, 1973. xvi, 104 p.

Examines present and future environmental concerns associated with the generation of electrical power. Cost considerations are included for both fossil fuel and nuclear plants along with possible technological improvements in the generation, distribution and transmission of electricity. Expenses are examined to determine the effect of pollution control on the cost of electric power.

Stepp, James M., and Macaulay, H.H. THE POLLUTION PROBLEM. Special Analysis no. 16. Washington, D.C.: American Enterprise Institute for Public Policy Research, 1968. 67 p.

General essay on the economic questions involved in pollution control, primarily water. Good, nontechnical discussion.

U.S. Comptroller General. WATER POLLUTION ABATEMENT PROGRAM: ASSESSMENT OF FEDERAL AND STATE ENFORCEMENT EFFORTS. Washington, D.C.: U.S. Environmental Protection Agency, 1972. 55 p.

Compilation of federal and state enforcement programs and legislative changes being considered by Congress.

U.S. Congress. House. Committee on Public Works. LAWS OF THE UNITED STATES RELATING TO WATER POLLUTION CONTROL AND ENVIRONMENTAL QUALITY. 91st Cong., 2d sess., Committee Print. Washington, D.C.: U.S. Government Printing Office, 1970. 265 p.

Updated compilation of legislation in the field of water pollution control and environmental quality. Includes a brief outline of each article of legislation followed by the actual law and/or amendments.

U.S. Congress. Senate. Committee on Public Works. THE ECONOMICS OF CLEAN WATER: SUMMARY REPORT OF THE FEDERAL WATER POLLUTION CONTROL ADMINISTRATION. 91st Cong., 2d sess., Committee Print. Washington, D.C.: U.S. Government Printing Office, 1970. 20 p.

Presents estimates of investment levels necessary to attain applicable water quality levels.

[U.S. Department of Interior.] U.S. Federal Water Pollution Control Administration. THE COST OF CLEAN WATER: ECONOMIC IMPACT ON AFFECTED UNITS OF GOVERNMENT. Washington, D.C.: 1968. 227 p.

Report of a study which evaluates the requirements of state and local governments to achieve desired water quality standards. It appraises the major financial and legal problems which will be encountered in implementing these standards.

______. FEDERAL WATER POLLUTION CONTROL ACT, PUBLIC LAW 660. Washington, D.C.: 1966. 19 p.

Complete copy of the act as amended.

U.S. Environmental Protection Agency. THE ECONOMICS OF CLEAN WATER. 4 vols. Washington, D.C.: 1972. Distributed by U.S. Government Printing Office.

Detailed account of the costs involved in cleaning up polluted waters; includes basic data used for the evaluation. Volume 1 describes the analysis and surveys which were undertaken to arrive at current investment estimates, volume 2 includes the methodology and data used for determining the pollution index, volume 3 reports industry expenditure for water pollution abatement, volume 4 summarizes the data and given estimates of capital investment necessary for water pollution abatement.

VanTassel, Alfred J., ed. ENVIRONMENTAL SIDE EFFECTS OF RISING INDUSTRIAL OUTPUT. Lexington, Mass.: Heath Lexington, 1970. xxiv, 548 p.

Collection of articles appraising pollution problems in the United States in the next thirty years from an economic point of view. Various areas such as industrial wastewater, electrical power generation, population growth, and pesticides are examined for their potential affect on pollution in the year 2000. Each author estimates current pollution coefficients (ratios of pollution to economic output) in various sectors and technologies and forecasts how these might charge.

Wenner, Lettie M. ATTITUDES OF WATER POLLUTION CONTROL OFFICIALS. DeKalb: Center for Governmental Studies, Northern Illinois University, 1973. 34 p.

Results of a questionnaire type evaluation of attitudes of federal, state, and local officials.

Wollman, Nathaniel, and Bonem, Gilbert W. THE OUTLOOK FOR WATER: QUALITY, QUANTITY, AND NATIONAL GROWTH. Resources for the Future. Baltimore: Johns Hopkins Press, 1971. xviii, 286 p.

Based on regional analysis, a systematic economic model is constructed for a national water perspective. Presents statistics, water use, treatment, wasteloads, and costs. Good source of statistics.

Zwick, David, and Benstock, Marcy. WATER WASTELAND. New York: Grossman Publishers, 1971. xvii, 494 p.

Report by Ralph Nader's study group on water pollution, the politics and administrative failures. Analyzes legislation, enforcement policies and the reasons for their failure.

C. JOURNAL ARTICLES

Blazey, D.R. "Legal Perspectives on Water Quality Management." A.S.C.E. JOURNAL HYDRAULICS DIVISION 101 (May 1975): 595-601.

Bruvold, W.H. "Human Perception and Evaluation of Water Quality." CRC CRITICAL REVIEWS IN ENVIRONMENTAL CONTROL 5 (March 1975): 153-231.

Bundgaard-Nielsen, M. "A Note on the Use of Economic Incentives in Water Pollution Abatement." WATER RESOURCES BULLETIN 12 (August 1976): 755-58.

Cross, F.L., Jr. "208 and You." POLLUTION ENGINEERING 8 (February 1976): 34-35.

Hipel, K.W.; Ragade, R.K.; and Unny, T.E. "Political Resolution of Environmental Conflicts." WATER RESOURCES BULLETIN 12 (August 1976): 813-27.

Marshall, H.E., and Ruegg, R.T. "Efficiency Problems from User Fees in Municipal Wastewater Treatment." WATER RESOURCES BULLETIN 12 (October 1976): 903-17.

Pisano, M. "Nonpoint Pollution: An EPA View of Areawide Water Quality Management." JOURNAL SOIL AND WATER CONSERVATION 31 (May-June 1976): 94-100.

Priscoli, J.D. "Citizen Advisory Groups and Conflict Resolution in Regional Water Resources Planning." WATER RESOURCES BULLETIN 11 (December 1975): 1233-43.

Schwarz, H.E. "Trouble in Paradise." A.S.C.E. JOURNAL HYDRAULICS DIVISION 101 (May 1975): 603-12.

Chapter 19
CASE STUDIES

A. ABSTRACTS AND BIBLIOGRAPHIES

[U.S. Department of Interior.] Water Resources Scientific Information Center. LAKE ERIE: A BIBLIOGRAPHY. Bibliography Series, WRSIC 72-209. Washington, D.C.: 1972. Distributed by National Technical Information Service. 240 p.

> Abstracts from SELECTED WATER RESOURCES ABSTRACTS through 15 May 1972 pertaining to Lake Erie.

______. LAKE HURON: A BIBLIOGRAPHY. Bibliography Series, WRSIC 72-210. Washington, D.C.: 1972. Distributed by National Technical Information Service. 95 p.

> Abstracts from SELECTED WATER RESOURCES ABSTRACTS on water quality and water pollution in Lake Huron, cover the period up to June 1975.

______. LAKE ONTARIO: A BIBLIOGRAPHY. Bibliography Series, WRSIC 72-212. Washington, D.C.: 1972. Distributed by National Technical Information Service. 200 p.

> Abstracts from SELECTED WATER RESOURCES ABSTRACTS on the water quality and pollution in Lake Ontario cover the period up to May 1972.

______. LAKE SUPERIOR: A BIBLIOGRAPHY. Bibliography Series, WRSIC 72-213. Washington, D.C.: 1972. Distributed by National Technical Information Service. 127 p.

> Abstracts from SELECTED WATER RESOURCES ABSTRACTS on Lake Superior cover the period ending June 1972.

______. LEGAL ASPECTS OF WATER POLLUTION IN NEW JERSEY AND PENNSYLVANIA: A BIBLIOGRAPHY. Bibliography Series, WRSIC 72-217. Washington, D.C.: 1972. Distributed by National Technical Information Service. iv, 223 p.

Abstracts from SELECTED WATER RESOURCES ABSTRACTS cover the period up to October 1972.

B. REFERENCES

Cleary, Edward J. THE ORSANCO STORY: WATER QUALITY MANAGEMENT IN THE OHIO VALLEY UNDER AN INTERSTATE COMPACT. Baltimore: Johns Hopkins Press, 1967. xvi, 335 p.

Details the Ohio River Valley Water Sanitation Commission from its inception. Describes the problems facing ORSANCO and their successes and failures.

Deininger, Rolf A., ed. MODELS FOR ENVIRONMENTAL POLLUTION CONTROL. Ann Arbor, Mich.: Ann Arbor Science Publishers, 1973. xi, 448 p.

Papers presented at a conference on "Systems Analysis for Environmental Pollution Control" deal with mathematical models and optimization techniques useful in pollution control including water pollution. Technical.

Dorfman, Robert; Jacoby, Henry D.; and Thomas, Harold A. MODELS FOR MANAGING REGIONAL WATER QUALITY. Cambridge, Mass.: Harvard University Press, 1972. 453 p.

Contains nine essays on the various analytical aspects of the problem of controlling water quality. Emphasis is on the development of analytical methods that encompass the technical and social facets of the water pollution problem.

Espey, W.H., Jr., and Ward, G.H., Jr. "Estuarine Water Quality Models." WATER RESEARCH 6 (October 1972): 1117-36.

Evaluation of the utility of models for simulating pollution in estuaries.

Falkner, Charles H. INSTITUTIONAL DESIGN FOR WATER QUALITY MANAGEMENT: A CASE STUDY OF THE WISCONSIN RIVER BASIN. Technical Report OWRRC - 1228. Madison: University of Wisconsin Water Resources Center, 1970. 300 p.

Technical report covering water quality management modeling and supporting studies. Contents include: dissolved oxygen prediction model, mathematical models in water quality management, temperature prediction model, and linear programming in water quality. Technical.

Fallows, James M. THE WATER LORDS. New York: Grossman Publishers, 1971. xxi, 294 p.

Details the study by Ralph Nader's group on the Savannah River, its pollution problems and the confrontation with industry. Social and economic aspects of water pollution are examined.

Gleeson, George W. RETURN OF A RIVER: THE WILLAMETTE RIVER. Water Resources Research Institute, WRRI-13. Corvallis: Oregon State University, 1972. 103 p.

Semihistorical description of the campaign to clean up the Willamette River.

Hubschman, J.H. "Lake Erie: Pollution Abatement, Then What?" SCIENCE 171 (12 February 1971): 536-40.

Effect of various pollutants on Lake Erie and its biota is discussed.

Ketelle, Martha J., and Uttormark, Paul D. PROBLEM LAKES IN THE UNITED STATES. Water Pollution Control Research Series 16010 EHR 12/71. Washington, D.C.: U.S. Environmental Protection Agency, Office of Research and Monitoring, 1971. iii, 282 p.

Compilation of statistics on 452 lakes in the United States. Information includes: location, size, depth, description of problem, remedial action being taken, and a description of available background data. Includes those lakes which have deteriorated to the extent rehabilitation is desirable.

McNulty, J. Kneeland. EFFECTS OF ABATEMENT OF DOMESTIC SEWAGE POLLUTION ON THE BENTHOS, VOLUMES OF ZOOPLANKTON AND THE FOULING ORGANISMS OF BISCAYNE BAY, FLORIDA. Coral Gables, Fla.: University of Miami Press, 1970. 107 p.

Text is primarily a comparison of conditions of the benthos, and so forth, before and after abatement of pollution. Includes a summary of the literature on pollution and the marine environment.

Minshall, G. Wayne, and Andrews, Douglas A. "An Ecological Investigation of the Portneuf River, Idaho: A Semi-Arid-Land Stream Subjected to Pollution." FRESHWATER BIOLOGY 3 (February 1973): 1-30.

Results of a study which assessed selected physical, chemical and biological conditions of the entire river over an extended period of time. Stream was used for irrigation and was the receptable of runoff from agricultural lands, wastes from a sewage treatment plant and wastes from phosphate processing operation. Major changes in the river were related to siltation from poor soil conservation practices and to waste discharge from the phosphate processing plant.

Regier, H.A., and Hartman, W.L. "Lake Erie's Fish Community: 150 Years of Cultural Stresses." SCIENCE 180 (22 June 1973): 1248-55.

Changes in Lake Erie due to man's activities are discussed including the effect of cultural stresses such as fisheries.

Van Tassel, Alfred J., ed. OUR ENVIRONMENT: THE OUTLOOK FOR 1980. Lexington, Mass.: Lexington Books, 1973. vii, 589 p.

One-half of the book is devoted to water pollution and examines various areas such as San Francisco Bay, Lake Michigan, Lower Mississippi River, as to the types of pollutants present, expenditures for pollution control, future plans, and projections for the area.

Part III

MISCELLANEOUS

Chapter 20

GUIDES TO BOOKS, BOOK REVIEWS, AND SCIENTIFIC MEETINGS

A.A.A.S. SCIENCE BOOKS: A QUARTERLY REVIEW. Washington, D.C.: American Association for the Advancement of Science, 1965-- . Quarterly.

Contains reviews of trade books, textbooks, and reference works in the pure and applied sciences, appraises books as to level of difficulty.

BOOKS IN PRINT 1979-80. 4 vols. New York: R.R. Bowker, 1979.

Provides title, author, date, publisher, price, series, edition, and other pertinent information on books published in the United States. Also published are FORTHCOMING BOOKS, BOOKS IN PRINT SUPPLEMENT AND PUBLISHERS WEEKLY, SUBJECT GUIDE TO FORTHCOMING BOOKS, and PAPERBOUND BOOKS IN PRINT--all designed to provide information on books recently published and those to be published in the future.

SCIENTIFIC MEETINGS. New York: Special Libraries Association, 1957-- . Quarterly.

Contains address and name of sponsoring organization, date, and place of conference, as well as costs, publications, and person to contact.

SERIAL SOURCES FOR THE BIOSIS DATA BASE. Philadelphia: Biosciences Information Service, 1978. Annual. Replaces BIOSIS LIST OF SERIALS.

Listing of serial literature of the life sciences, including review annuals, reports of periodically published conferences, congresses and symposia, and government-sponsored research reports, as well as traditional journal literature. Individual entries include full title, abbreviated title, frequency of publication and the name and address of the publisher. Indicates those which have ceased publication or changed titles.

TECHNICAL BOOK REVIEW INDEX. New York: Special Libraries Association/ Monthly, 1935-- . Monthly.

> Compilation of reviews of scientific and technical books found in scientific, technical and trade journals. Coverage spans the pure sciences to engineering, medicine, and related subjects.

WORLD MEETINGS: UNITED STATES AND CANADA. New York: Macmillan Information, 1968-- . Quarterly.

> Two-year registry of future medical, scientific, and technical meetings. Citation indicates sponsor, title, place of the conference, date, coverage, and deadlines. Includes the address for obtaining information on those attending and the form and title of any publication of the meeting.

Chapter 21

GUIDES TO ENVIRONMENTAL LEGISLATION

CODE OF FEDERAL REGULATIONS (CFR). Washington, D.C.: Office of Federal Register, National Archives and Records Service, General Services Administration. 1938-- . Annual.

> CFR is a "codification" of the general and permanent rules published in the FEDERAL REGISTER by the executive departments and agencies of the federal government. CFR is kept up to date by individual issues of the FEDERAL REGISTER. Sections pertinent to water pollution include: chapter 18, "Conservation of Power and Water Resources," chapter 40, "Protection of Environment," chapter 50, "Wildlife and Fisheries."

ENVIRONMENT REGULATION HANDBOOK. New York: Environment Information Center, 1973-- . Monthly.

> Contains information on regulations, an official directory, citations of state environmental laws and a regulatory bibliography.

ENVIRONMENT REPORTER. Washington, D.C.: Bureau of National Affairs, Inc., 1970-- . Weekly.

> Covers current legislation, federal and state regulations, court decisions and developments dealing with the environment. Besides the weekly update, monographs are published at irregular intervals which present an overview of a particular topic such as thermal pollution. Good source of information for both federal and state decisions and laws.

UNITED NATIONS YEARBOOK. New York: Office of Public Information, United Nations, 1946-- . Annual.

> Presents an account of United Nations activities in areas ranging from politics to the environment.

Chapter 22

GUIDES TO CAREERS AND RESEARCH INSTITUTIONS

A. CAREERS

Fanning, Odom. OPPORTUNITIES IN ENVIRONMENTAL CAREERS. New York: Vocational Guidance Manuals, 1971. 271 p.

Book indicates the disciplines which will be required in the next decade to deal with environmental problems. Analysis is made of manpower needs and projected job opportunities in environmental management.

Hopke, William E., ed. THE ENCYCLOPEDIA OF CAREERS AND VOCATIONAL GUIDANCE. 3d rev. ed. 2 vols. Chicago: J.G. Ferguson Publishing, 1975. Distributed by Doubleday.

Presents descriptions of many different careers, what they encompass, opportunities and the type of preparation needed.

U.S. Department of Labor. Bureau of Labor Statistics. OCCUPATIONAL OUTLOOK HANDBOOK 1974-75 EDITION. Bulletin 1785. Washington, D.C.: U.S. Government Printing Office, 1949-- . Biennial.

Presents general job descriptions, nature of the work, places of employment, training, outlook, earnings, and working conditions, plus where to obtain additional sources of information.

B. RESEARCH INSTITUTIONS

Listings of research institutions are contained in numerous directories which have already been cited. Therefore, only the bibliographic information will be repeated. The reader is referred to chapter 7, "Directories," for the annotations.

DIRECTORY OF INFORMATION RESOURCES IN THE UNITED STATES: WATER. Washington, D.C.: U.S. Library of Congress, National Referral Center, 1966. Distributed by U.S. Government Printing Office. v, 248 p.

DIRECTORY OF WATER POLLUTION RESEARCH LABORATORIES. Publication no. 18313. Paris: Organization for Economic Cooperation and Development, 1966. 519 p.

Onyx Group, Inc. ENVIRONMENT U.S.A.: A GUIDE TO AGENCIES, PEOPLE AND RESOURCES. New York: R.R. Bowker, 1974. xii, 451 p.

Syracuse University Research Corporation Policy Institute. ENVIRONMENTAL RESEARCH LABORATORIES IN THE FEDERAL GOVERNMENT: AN INVENTORY. 2 vols. Syracuse, N.Y.: 1971.

Thibeau, C.E., ed. N.F.E.C. DIRECTORY OF ENVIRONMENTAL INFORMATION SOURCES. 2d ed. Boston: National Foundation for Environmental Control, 1972. Distributed by Canners Books. ix, 457 p.

Wilson, William K.; Dowd, Morgan D.; and Sholtys, Phyllis A., eds. WORLD DIRECTORY OF ENVIRONMENTAL RESEARCH CENTERS. 2d ed. New York: Onyx Press, 1974. Distributed by R.R. Bowker. xi, 330 p.

Chapter 23

EDUCATIONAL PROGRAMS IN WATER POLLUTION

The study of water pollution has traditionally been concentrated in the department of civil engineering. Today the various facets of water pollution are studied in the departments of chemistry, botany, biology and zoology, and medicine as well as in the departments of economics, sociology, and law. In addition, there has been a proliferation of environmental programs, some interdisciplinary, leading to degrees in environmental studies, environmental engineering, and ecology.

The list of colleges is restricted to those offering a bachelor's degree or higher in programs which concentrate on the study of water pollution. The reader is reminded that while many schools do not offer specific programs in the study of water pollution and/or the environment, this area may be covered within traditional programs in engineering, biology, economics, and so forth.

A. DIRECTORIES

Agricultural Sciences Information Network. DIRECTORY OF INFORMATION RESOURCES IN AGRICULTURE AND BIOLOGY. Beltsville, Md.: National Agricultural Library, 1971. Distributed by U.S. Government Printing Office. 523 p.

> Listing of land grant colleges indicating the staff, scope of courses, research areas, and publications, plus funding and degrees offered.

American Institute of Biological Sciences. DIRECTORY OF BIOSCIENCE DEPARTMENTS IN THE UNITED STATES AND CANADA. New York: Reinhold Publishing, 1967. 672 p.

> Accredited academic institutions are listed with information pertaining to courses offered, requirements, composition of departments, and if they have field stations.

DIRECTORY OF INFORMATION RESOURCES IN THE UNITED STATES: BIOLOGICAL SCIENCES. Washington, D.C.: Library of Congress, National Referral Center, 1972. Distributed by U.S. Government Printing Office. iv, 577 p.

Title, addresses, area of interest, and publications of universities with programs in the biological sciences.

DIRECTORY OF INFORMATION RESOURCES IN THE UNITED STATES: WATER. Washington, D.C.: Library of Congress, National Referral Center, 1966. Distributed by U.S. Government Printing Office. v, 248 p.

Describes numerous organizations and institutions which are conducting research or collecting data on water and water related subjects. Focuses primarily on freshwater.

Quigg, Philip W., ed. WORLD DIRECTORY OF ENVIRONMENTAL EDUCATION PROGRAMS: POST-SECONDARY STUDY AND TRAINING IN 70 COUNTRIES. New York: R.R. Bowker, 1973. xiii, 289 p.

Addresses of institution, cost of program, scholarships, and description of areas covered are given for more than one thousand programs.

B. COLLEGES AND UNIVERSITIES

The reference includes address of institution, programs, and degrees offered. For more information on the following references, refer to Quigg, WORLD DIRECTORY OF ENVIRONMENTAL EDUCATION PROGRAMS, see entry above.

ALABAMA

Alabama Agricultural and Mechanical University
Normal, Ala. 35762

Natural Resources and Environmental Studies: BS

Auburn University
Auburn, Ala. 36830

Fisheries and Allied Aquacultures: BS, MS, Ph.D

Tuskegee Institute
Tuskegee, Ala. 36088

Environmental Sciences: MS

University of Alabama
University, Ala. 35486

Graduate Training in Sanitary Engineering: MS, Ph.D

ALASKA

University of Alaska
College, Alaska 99735

Environmental Health Engineering Program: MS

ARIZONA

University of Arizona
Tucson, Ariz. 85271

Hydrology and Water Resources Administration: BS, MS, Ph.D
Sanitary Engineering: BS, MS, Ph.D

ARKANSAS

University of Arkansas
Fayetteville, Ark. 72701

Environmental Science: BS
Environmental Engineering, Graduate Studies: MS, Ph.D

CALIFORNIA

California Institute of Technology
1201 East California Boulevard
Pasadena, Calif. 91109

Environmental Engineering Science: MS, Ph.D
Environmental Quality Laboratory

California Polytechnic State University
San Luis Obispo, Calif. 93401

Environmental Engineering: BS, Master of Engineering
Natural Resources Management: BS, MS

California State University, Fullerton
Fullerton, Calif. 92634

Environmental Studies Program: MS

Claremont Graduate School
Ninth and College Avenue
Claremont, Calif. 91711

Government: Ph.D

University of California, Berkeley
Berkeley, Calif. 94720

Group Major in Environmental Studies: BS

University of California, Davis
Davis, Calif. 95616

Ecology Group: MS, Ph.D

University of California, Irvine
Irvine, Calif. 92664

Civil and Environmental Engineering: BS, MS, Ph.D

University of California, Los Angeles
405 Hilgard Avenue
Los Angeles, Calif. 90024

Environmental Science and Engineering: Ph.D

University of California, Santa Barbara
Santa Barbara, Calif. 93106

Environmental Studies: BS

University of California, Scripps Institution of Oceanography
Box 109
La Jolla, Calif. 92037

Applied Ocean Science: MS, Ph.D

University of Southern California
Los Angeles, Calif. 90007

Environmental Engineering: BS, MS, Ph.D
Center for Urban Affairs: BA, Ph.D

COLORADO

Colorado School of Mines
Guggenheim Hall
Golden, Colo. 80401

Mineral Engineering: BS, MS, Ph.D

Colorado State University
Fort Collins, Colo. 80521

Environmental Resources Center: BS, MS, Ph.D

University of Colorado
Boulder, Colo. 80302

Environmental Conservation: BS

University of Denver
University Park
Denver, Colo. 80210

Environmental Engineering: BS, MS, Ph.D.

CONNECTICUT

Sacred Heart University
5229 Park Avenue
Bridgeport, Conn. 06604

Yale University
204 Prospect Street
New Haven, Conn. 06511

School of Forestry and Environmental Studies: MS, Ph.D.

DELAWARE

University of Delaware
Newark, Del. 19711

Environmental Engineering: M Eng., Ph.D
College of Marine Studies: MS, Ph.D

DISTRICT OF COLUMBIA

George Washington University
Washington, D.C. 20006

Environmental Engineering: BS, MS

Howard University
2300 Sixth Street, N.W.
Washington, D.C. 20001

Bioenvironmental Engineering and Sciences: BS, MS, Ph.D

FLORIDA

Florida Institute of Technology
Country Club Road
Melbourne, Fla. 32901

Environmental Biology-Environmental Studies: BS, MS

Florida State University
Tallahassee, Fla. 32306

Environmental Studies Concentration in Social Science Major: BS

Florida Technological University
P.O. Box 25000
Orlando, Fla. 32816

Environmental Engineering: BS
Environmental Systems Engineering: MS

University of Florida
Gainesville, Fla. 32601

Environmental Engineering Sciences: BS, MS, Ph.D

University of West Florida
Pensacola, Fla. 32504

Environmental Studies: BA

GEORGIA

Georgia Institute of Technology
Atlanta, Ga. 30332

Department of City Planning: Master's in City Planning, Ph.D

University of Georgia
Athens, Ga. 30601

Ecology: Ph.D.
Water Resources Program: BS, MS, Ph.D
Environmental Health Science: BS

HAWAII

University of Hawaii
Honolulu, Hawaii 96822

School of Public Health: BA

IDAHO

College of Idaho
Caldwell, Idaho 83605

Snake River Regional Studies Center

Idaho State University
Pocatello, Idaho 83201

Center for Environmental Studies

ILLINOIS

Concordia Teachers College
7400 Augusta
River Forest, Ill. 60305

Concentration in Environmental Studies: BA

DePaul University
2323 North Seminary Avenue
Chicago, Ill. 60614

Environmental Studies Courses: BA, BS, MA, Ph.D

Illinois Institute of Technology
3300 South Federal Streeet
Chicago, Ill. 60616

Department of Environmental Engineering: MS, Ph.D

Northeastern Illinois University
Bryn Mawr at St. Louis Avenues
Chicago, Ill. 60625

Geography and Environmental Studies: BA

University of Illinois, Urbana-Champaign
Urbana, Ill. 61801

Environmental Studies Program: MS, Ph.D

University of Illinois Medical Center
School of Public Health
P.O. Box 6998
Chicago, Ill. 60680

Environmental Health Sciences: MS, Ph.D

INDIANA

Ball State University
Muncie, Ind. 47306

Natural Resources: BS, BA, MS, MA

Indiana State University
Terre Haute, Ind. 47809

Ecology: MA, Ph.D

Indiana University
Bloomington, Ind. 47401

Environmental Policy and Management: Bachelor and Masters's

Manchester College
604 College Avenue
North Manchester, Ind. 46962

Environmental Studies: BS, BA

Purdue University
West Lafayette, Ind. 47907

Institute for Environmental Health: BS, MS, Ph.D
Environmental Engineering: BS, MS, Ph.D

Rose-Hulman Institute of Technology
5500 Wabash Avenue
Terre Haute, Ind. 47803

Biological Engineering: BS, MS

University of Notre Dame
Notre Dame, Ind. 46556

Environmental Health Engineering: MS, Ph.D
Environmental Research Center

IOWA

Central College
University at Broadway
Pella, Iowa 50219

Interdisciplinary Programs: BA

Iowa State University
Ames, Iowa 50010

Water Resources: MS, Ph.D

University of Iowa
Iowa City, Iowa 52240

Urban and Regional Planning: MA, MS

KANSAS

Kansas State University
Manhattan, Kan. 66506

Regional and Community Planning: MS
Engineering Studies: BS, MS, Ph.D

Saint Mary of the Plains College
Dodge City, Kans. 67801

Environmental Pollution: BA

University of Kansas
Lawrence, Kans. 66044

Environmental Health, Engireering and Science: MS, Ph.D
Water Resources Engineering and Science: MS

KENTUCKY

Eastern Kentucky University
Richmond, Ky. 40475

Regional Planning: BA

Western Kentucky University
Bowling Green, Ky. 42101

Environmental Engineering Technology: BS
Environmental Health: BS

LOUISIANA

Louisiana State University
Room 304, Field House
Baton Rouge, La. 70803

School Environmental Design: BA, MA

Louisiana Tech University
Tech Station
Ruston, La. 71270

Environmental Protection Agency, Water Quality Training: MS, M Eng.
Water Resources Center

McNeese State University
4100 Ryan Street
Lake Charles, La. 70601

Environmental Science: BS, MS

Tulane University
6823 St. Charles Avenue
New Orleans, La. 70118

Environmental Engineering: MS, M Eng., Ph.D, Eng. D

MAINE

Bowdoin College
Brunswick, Maine 04011

Environmental Studies Program: BA
Research Projects in Avian Ecology

University of Maine
Orono, Maine 04473

Natural Resource Management: BS

MARYLAND

Johns Hopkins University
Baltimore, Md. 21218

Environmental Engineering: MS, MA, Ph.D

University of Maryland
College Park, Md. 20740

Environmental Management Project

MASSACHUSETTS

Boston University
Boston, Mass. 02215

Environmental Studies: BA, BS

Framingham State College
Framingham Centre, Mass. 01701

Environmental Studies Emphasis in Biology: BS

Harvard University
Cambridge, Mass. 02138

Environmental Sciences and Engineering: BS, BS, MS, Ph.D
Environmental Systems Program: Ph.D

Massachusetts Institute of Technology
Cambridge, Mass. 02139

Environmental Management: BS, MS, Ph.D
Environmental Concentration: BS

Newton College of the Sacred Heart
885 Centre Street
Newton, Mass. 02159

Urban Studies: BA

University of Massachusetts at Amherst
Amherst, Mass. 01002

Environmental Engineering Program: MS, Ph.D

Woods Hole Oceanographic Institution
Woods Hole, Mass. 02543

Oceanography/Ocean Engineering: Ph.D, Sc.D

MICHIGAN

General Motors Institute
1700 West Third Avenue
Flint, Mich. 48502

Environmental Engineering, Bachelor's

Hope College
Holland, Mich. 49423

Institute for Environmental Quality: BA

Michigan State University
East Lansing, Mich. 48223

Ecological Psychology: MA, Ph.D

Michigan Technological University
Houghton, Mich. 49931

Environmental Engineering: BS, MS
Environmental Biology: BS, MS, Ph.D

University of Michigan
Ann Arbor, Mich. 48104

Environmental Science Engineering: BS
University Program in Water Resources: MS, Ph.D

MINNESOTA

Bemidji State College
Fourteenth and Birchmont
Bemidji, Minn. 56601

Center for Environmental Studies: BS, BA

Concordia College
Moorehead, Minn. 56560

Environmental Minor: BA, BS

University of Minnesota at Duluth
Duluth, Minn. 55812

Environmental Courses: BA, BS, MA, MS

University of Minnesota at Minneapolis
Twin Cities Campus
Minneapolis, Minn. 55455

Environmentally-Oriented Engineering Program: BS
Environmentally Related Programs: BA, BS
Limnological Research Center: MS, Ph.D

University of Minnesota at St. Paul
Twin Cities Campus
St. Paul, Minn. 55101

Environmentally-Oriented Programs: BS, MS, Ph.D

MISSISSIPPI

Mississippi State University
P.O. Drawer 6H
State College, Miss. 39762

Institute for Environmental Studies

University of Southern Mississippi
Southern Station, Box 1
Hattiesburg, Miss. 39401

Environmental Science Degree Program: BS

MISSOURI

Lindenwood Colleges
St. Charles, Mo. 63301

Environmental Biology: BA

Southwest Missouri State College
Springfield, Mo. 65802

Environmental Chemical Technology: BS

University of Missouri at Rolla
Rolla, Mo. 65401

School of Engineering: BS, MS, Ph.D

MONTANA

Montana College of Mineral Science and Technology
West Park Street
Butte, Mont. 59701

Environmental Engineering: BS, MS

Montana State University
Bozeman, Mont. 59715

Environmental Health Engineering: MS, Ph.D

NEBRASKA

Dana College
Blair, Nebr. 68008

Environmental Studies: BS

University of Nebraska
Lincoln, Nebr. 68508

Natural Resources Options: BS

NEVADA

University of Nevada-Reno
Reno, Nev. 89507

Environmental Science

NEW HAMPSHIRE

Plymouth State College of the University of New Hampshire
Plymouth, New Hampshire 03264

Programs offered through various departments: BA, BS

University of New Hampshire
Durham, H.H. 03824

Institute of Natural and Environmental Resources: BS, MS
Marine Science and Ocean Engineering: BS, MS, Ph.D

NEW JERSEY

Newark College of Engineering
323 High Street
Newark, N.J. 07102

Industrial Waste Water Control Program: MS

Princeton University
Princeton, N.J. 08540

Center for Environmental Studies: interdisciplinary degrees from various departments

Rutgers University--Cook College
New Brunswick, N.J. 08903

Department of Environmental Sciences: MS, Ph.D

Rutgers University--State University of New Jersey
New Brunswick, N.J. 08903

Environmental Engineering: MS, Ph.D
Institute for Environmental Studies: BS, MS, Ph.D

NEW MEXICO

Eastern New Mexico University
Portales, N. Mex. 88130

Environmental Health: BS

New Mexico Institute of Mining and Technology
Campus Station
Socorro, N. Mex. 87801

Environmental Engineering: BS

NEW YORK

Barnard College--Columbia University
New York, N.Y. 10027

Environmental Conservation and Management: BA

The City College--City University of New York
138th Street and Convent Avenue
New York, N.Y. 10031

Department of Civil Engineering: B Eng., M Eng., Ph.D

Columbia University
New York, N.Y. 10027

Environmental Science and Engineering: MS, Eng. Sc.D, Ph.D

Cornell University
Ithaca, N.Y. 14850

College of Agriculture and Life Sciences: BS, MS, Ph.D
Environmental Studies Programs: interdisciplinary with degrees offered through several departments

Elmira College
Elmira, N.Y. 14901

Environmental Studies: BA, BS

Fordham University
Louis Calder Conservation and Ecology Study Center
31 Whippoorwill Road
Armonk, N.Y. 10504

Environmental Science: BA, BS, MS, Ph.D

Iona College
North Avenue
New Rochelle, N.Y. 10801

Environmental Studies Major: BS

Manhattan College
Manhattan College Parkway
Riverdale, N.Y. 10471

Environmental Engineering and Science: M Eng., MS
Environmental Studies
Institute of Aquatic Biology: BS

Marist College
North Road
Poughkeepsie, N.Y. 12603

Environmental Science Program: BA

New York University
100 Washington Square East
New York, N.Y. 10003

Environmental Health Science: MS

Rensselaer Polytechnic Institute
Eighth Street
Troy, N.Y. 12181

Environmental Engineering: E Eng., MS, M Eng., Ph.D., Eng. D

State University of New York at Buffalo
Buffalo, N.Y. 14214

Center for Inland Water Resources: Ph.D

State University of New York at Stony Brook
Stony Brook, N.Y. 11790

Marine Environmental Studies Program: MS

State University of New York College at Oneonta
Oneonta, N.Y. 13820

Biology Department: BA, BS, MA

Syracuse University
Syracuse, N.Y. 13210

Environmental Engineering Program: BS

University of Rochester
College of Arts and Sciences
River Campus Station
Rochester, N.Y. 14627

Biology and Geology: BS

NORTH CAROLINA

East Carolina University
Greenville, N.C. 27834

Department of Environmental Health: BS

University of North Carolina
Chapel Hill, N.C. 27514

Department of Environmental Sciences and Engineering: MS, Ph.D
UNC Wastewater Research Center: MS, Ph.D

NORTH DAKOTA

University of North Dakota
University Station
Grand Forks, N.C. 58201

Institute for Ecological Studies: BS, MS, Ph.D

OHIO

Antioch College
Yellow Springs, Ohio 45387

Environmental Studies Center: BA

Cleveland State University
Euclid Avenue and 24th Street
Cleveland, Ohio 44115

Environmental Chemistry: MS

Institute for Environmental Education
8911 Euclid Avenue
Cleveland, Ohio 44160

Cuyahoga River Watershed Prcject: MS

University of Cincinnati
Cincinnati, Ohio 45221

Environmental Engineering and Science: BS, MS, Ph.D

University of Toledo
2801 West Bancroft Street
Toledo, Ohio 43606

Department of Engineering Env'ronmental Studies: MS, Ph.D

OKLAHOMA

University of Oklahoma
Norman, Okla. 73069

Civil Engineering and Environmental Sciences: BS, MS, Ph.D

OREGON

University of Oregon
Eugene, Oreg. 97403

Environmental Studies Center: BA, BS, MA, MS

PENNSYLVANIA

California State College
Third Street
California, Pa. 15419

Environmental Studies: BA

Carnegie Mellon University
Schenley Park
Pittsburgh, Pa. 15212

Environmental Studies Institute: interdisciplinary degrees offered through participating departments

Dickinson College
Carlisle, Pa. 17013

Environmental Science: BA, BS

Drexel University
32d and Chestnut Streets
Philadelphia, Pa. 19104

Environmental Engineering and Science: MS

Lehigh University
Bethlehem, Pa. 18015

Environmental Science and Resource Management: BS

Mercyhurst College
501 East 38th Street
Erie, Pa. 16501

Environmental Education: BA

Muhlenberg College
Allentown, Pa. 18104

Environmental Studies Program: BS

Pennsylvania State University
University Park, Pa. 16802

Environmental Pollution Control: MS, ME

Slippery Rock State College
Slippery Rock, Pa. 16057

Environmental Education: BA, MS, M Ed.
Environmental Sciences: BA

University of Pittsburgh
Pittsburgh, Pa. 15213

Environmental Health Program: MS, DS
Public Works Administration: MS
Civil Engineering: BS, MS, Ph.D

PUERTO RICO

Universidad de Puerto Rico
Mayaguez, Puerto Rico 00708

Department of Marine Sciences: MS, Ph.D

RHODE ISLAND

University of Rhode Island
Kingston, R.I. 02881

Institute of Environmental Biology: BS, BA, MS, MA, Ph.D
Oceanography: MS, Ph.D

SOUTH CAROLINA

Clemson University
Clemson, S.C. 29631

Environmental Systems Engineering Department: MS, Ph.D
Water Resources Engineering: MS, Ph.D
Agricultural Engineering: BS, M Eng., MS, Ph.D

SOUTH DAKOTA

South Dakota State University
Brookings, S. Dak. 57006

Environmental Management

TENNESSEE

Christian Brothers College
650 East Parkway South
Memphis, Tenn. 38104

Water Pollution Analysis: BS

East Tennessee State University
Johnson City, Tenn. 37601

Environmental Health: BS, MS

University of Tennessee
Knoxville, Tenn. 37916

Water Resources Development: MS
Environmental Engineering Program: BS, MS, Ph.D

Vanderbilt University
Box 6304, Station B
Nashville, Tenn. 37235

Engineering and Water Resources Engineering: BS, BE, MS, Ph.D

TEXAS

Baylor University
Waco, Tex. 76703

Institute of Environmental Studies: BS, BA, MS

Lamar University
P. O. Box 10208
Beaumont, Tex. 77710

Environmental Studies: BS

North Texas State University
Box 5218, North Texas Station
Denton, Tex. 76203

Institute for Environmental Studies: MA, MS, Ph.D

Rice University
6100 South Main
Houston, Tex. 77001

Environmental Science and Engineering: BA, MS, Ph.D

Sam Houston University
Huntsville, Tex. 77430

Environmental Sciences: BS

Southern Methodist University
SMU, Box 945
Dallas, Tex. 75222

Environmental Systems: BS, MS, M Eng., Ph.D

Southwest Texas State University
San Marcos, Tex. 78666

Aquatic Studies Program: BS, MS

Texas A & M University
College of Marine Sciences and Maritime Resources
Marine Laboratory Building 311
Fort Crockett
Galveston, Tex. 77550

Marine Laboratory: MS, Ph.D

Texas Christian University
TCU Station
Fort Worth, Tex. 76129

Environmental Sciences: BS, MS

Texas Tech University
Lubbock, Tex. 79409

Agricultural Engineering: BS, MS, Ph.D

University of Houston
Cullen Boulevard
Houston, Tex. 77004

University of Houston Coastal Center: BS, MS, Ph.D

University of Texas at Austin
Austin, Tex. 78712

Environmental Health Engineering: MS, Ph.D

Wayland Baptist College
Plainview, Tex. 79072

Biological Sciences: BS

UTAH

University of Utah
Salt Lake City, Utah 84112

Environmental Studies Committee: BS, MS, Ph.D

VERMONT

Johnson State College
Johnson, Vt. 05656

Environmental Studies Program: BA, BS

Norwich University
Northfield, Vt. 05663

Engineering and Environmental Engineering: B Eng.

VIRGINIA

Old Dominion University
5215 Hampton Boulevard
Norfolk, Va. 23508

Environmental Health: BS

Virginia Institute of Marine Science
Gloucester Point, Va. 23062

Biological, General and Fisheries Oceanography: MA, Ph.D

WASHINGTON

University of Washington
Friday Harbor, Wash. 98250

Friday Harbor Laboratories: interdisciplinary

Washington State University
Pullman, Wash. 99163

Environmental Science Program: BS, MS

Whitworth College
Spokane, Wash. 99218

Environmental Studies: BA, BS

WEST VIRGINIA

West Virginia University
Morgantown, W. Va. 26506

Department of Biology: BA, MS, MA, Ph.D

WISCONSIN

Alverno College
3401 South 39th Street
Milwaukee, Wis. 53215

Environmental Studies: BS

Institute of Paper Chemistry
1043 East South River Street
Appleton, Wis. 54911

Institute of Paper Chemistry: MS, Ph.D

University of Wisconsin--Green Bay
Green Bay, Wis. 54302

College of Environmental Sciences: BA, BS

University of Wisconsin--Madison
Madison, Wis. 53706

Water Chemistry Program: MS, Ph.D

University of Wisconsin--Milwaukee
College of Applied Science and Engineering
Science Complex Building
Milwaukee, Wis. 53201

Water Resources Management: MS, M Eng., D. Eng.

University of Wisconsin--Whitewater
Whitewater, Wis. 53190

Resources of Geography: BS

Wisconsin State University--Oshkosh
800 Algoma Boulevard
Oshkosh, Wis. 54901

Biology Major-Environmental Option: BA, BS, MS

WYOMING

University of Wyoming
Laramie, Wyo. 82070

Water Resources: MS

Chapter 24
GRANT SOURCES

ANNUAL REGISTER OF GRANT SUPPORT: 1978-1979. 12th ed. Chicago: Marquis Academic Media, 1978. 736 p.

> Annual guide to grant support programs of government agencies, foundations, business and professional organizations. Includes fellowships. Indicates person to contact, amount of money available, and type of research sponsored.

The following is a list of sources of grants for research in water pollution. Please refer to the ANNUAL REGISTER OF GRANT SUPPORT for an updated and a more detailed citation.

American Water Works Association, 6666 West Quincy Avenue, Denver, Colo. 80234.

> Funding available through "Academic Achievement Award" which is awarded to the best graduate thesis on water supply subjects. Inquiries are sent to the above address.

Environmental Protection Agency, Office of Air and Waste Management, Washington, D.C. 20460.

> Office funds numerous programs related to the prevention and and control of air and solid waste pollution. Funds are also available for the development of career oriented personnel qualified to work in pollution abatement and control. Other sponsored programs include the demonstration of methods for solid waste disposal, recycling, or recovery of materials from waste. Inquiries should be sent to: Grants Administration Division, 401 M Street, S.W., Washington, D.C. 20460.

Environmental Protection Agency, Office of Planning and Management, Washington, D.C. 20460.

> Funding is available through two programs which come under the general heading, Environmental Protection Consolidated

Grants. Their purpose is to provide for coordination and management of environmental approaches to pollution problems, including water pollution control. Send inquiries to: Grants Administration Division, 401 M Street, S.W., Washington, D.C. 20460.

Environmental Protection Agency, Office of Research and Development, Washington, D.C. 20460.

Projects funded through this office include those relating to the causes, effects, extent, prevention and control of various types of pollution. Send inquiries to: Grants Administration Divsion, 401 M Street, S.W., Washington, D.C. 20460.

Environmental Protection Agency, Office of Water and Hazardous Materials, Washington, D.C. 20460.

Numerous programs are funded which deal with planning, implementing and improving environmental programs; also with the establishment and maintenance of programs for the prevention and control of water pollution. Funded programs include those to develop qualified personnel to work in pollution abatement and control and to provide the latest information available in the field of water pollution. Send inquiries to: Grants Administration Division, 401 M Street, S.W., Washington, D.C. 20460.

Ford Foundation, 320 East 43d Street, New York, N.Y. 10017.

Grants awarded through "Resources and the Environment" support ecological approaches in the education of research planners and administrators; support policy studies and projects which involve international cooperation. Send inquiries to above address.

National Environmental Health Association, 1200 Lincoln, Suite 704, Denver, Colo. 80203.

The "Undergraduate Scholarship Program" provides funds to students in the field of environmental health science. Send inquiries to: Nick Pohlit, Executive Director, National Environmental Health Association, 1200 Lincoln, Suite 704, Denver, Colo. 80203.

National Institute of Environmental Health Sciences, P.O. Box 12233, Research Triangle Park, N.C. 27709.

Grants are funded through the "Environmental Health Sciences Research Program" to support research into the source, distribution, mode of entry, and effects of environmental agents on biological systems. Send inquiries to: Chief, Scientific Programs Branch, Extra-mural Programs, National Institute of Environmental Health Sciences, P.O. Box 12233, Research Triangle Park, N.C. 27709.

National Research Council, Associateship Office (JH 606-Q), 2101 Constitution Avenue, N.W., Washington, D.C. 20418.

A postdoctoral and senior postdoctoral residence research associateship are sponsored through the "National Research Council - U.S. Army Medical Bioengineering Research and Development Laboratory Resident Research Associateship" for research and advanced study in all areas of pollution control including water pollution and pesticide disposal. Send inquiries to above address.

National Science Foundation, 1800 G Street, N.W., Washington, D.C. 20550.

Through "Research Applied to National Needs (RANN): Environment" research is supported in such areas as regional environmental management, environmental hazards, and chemical threats to man and the environment. Send inquiries to: Division of Advanced Environmental Research Technology, National Science Foundation, 1800 G Street, N.W., Washington, D.C. 20550.

National Science Foundation, 1800 G Street, N.W., Washington, D.C. 20550.

The program "Environmental Biology" supports research in such areas as ecology, ecosystem studies, systematic biology, and research resources. Send inquiries to Division of Envirormental Biology, National Science Foundation, 1800 G Street, N.W., Washington, D.C. 20550.

National Wildlife Federation, 1412 16th Street, N.W., Washington, D.C. 20036.

Fellowships are awarded through "Environmental Conservation Fellowships" to support study in the following areas related to water pollution: pollution control and abatement, soil conservation, ecosystem analysis, and modeling for resource management. Send inquiries to Executive Vice President, National Wildlife Federation, 1412 16th Street, N.W., Washington, D.C. 20036.

Office of Water Research and Technology, Department of the Interior, 18th and C Streets, N.W., Washington, D.C. 20240.

Basic and applied research is sponsored through the program "Water Research and Development" in any area of the physical, life and social sciences as they pertain to water and water-related resources. Send inquiries to above address.

Rockefeller Foundation, 1133 Avenue of the Americas, New York, N.Y. 10036.

Grants are given through "Quality of the Environment" to support research which will enhance the physical environment including water quality. Send inquiries to: The Secretary, Rockefeller Foundation, 1133 Avenue of the Americas, New York, N.Y. 10036.

Chapter 25

ADDRESSES OF FEDERAL, REGIONAL, AND STATE WATER AGENCIES

A. FEDERAL

U.S. Environmental Protection Agency
401 M Street, S.W.
Washington, D.C. 20460

Region I: Room 2302
John F. Kennedy Federal Building
Boston, Mass. 02203
(Connecticut, Maine, Massachusetts, New Hampshire, Rhode Island, Vermont)

Region II: Room 847
26 Federal Plaza
New York, N.Y. 10007
(New Jersey, New York, Puerto Rico, Virgin Islands)

Region III: Curtis Building
Sixth and Walnut Streets
Philadelphia, Pa. 19106
(Delaware, Maryland, Pennsylvania, Virginia, West Virginia, District of Columbia)

Region IV: Suite 300
1421 Peachtree Street, N.E.
Atlanta, Ga. 30309
(Alabama, Florida, Georgia, Kentucky, Mississippi, North Carolina, South Carolina, Tennessee)

Region V: 230 South Dearborn
Chicago, Ill. 60604
(Illinois, Indiana, Michigan, Minnesota, Ohio, Wisconsin)

Region VI: Suite 1100
1600 Patterson Street
Dallas, Tex. 75202
(Arkansas, Louisiana, New Mexico, Texas, Oklahoma)

Region VII: Room 249
1735 Baltimore Avenue
Kansas City, Mo. 64106
(Iowa, Kansas, Missouri, Nebraska)

Region VIII: Room 900
Lincoln Tower Building
1860 Lincoln Street
Denver, Colo. 80203
(Colorado, Montana, North Dakota, South Dakota, Utah, Wyoming)

Region IX: 100 California Street
San Francisco, Calif. 94111
(Arizona, California, Hawaii, Nevada; American Samoa, Guam, Trust Territories of Pacific Islands, Wake Island)

Region X: 1200 Sixth Avenue
Seattle, Wash. 98108
(Alaska, Idaho, Oregon, Washington)

B. REGIONAL

Delaware River Basin Commission
P.O. Box 7360
Trenton, N.J. 08628
(signatories: Delaware, New Jersey, Pennsylvania, United States)

Hudson River Valley Commission
c/o New York State Parks and Recreation
Empire State Plaza
South Swan Street Building
Albany, N.Y. 12223
(state: New York)

Interstate Commission on the Potomac River Basin
Suite 814, East West Towers
4350 East West Highway
Bethesda, Md. 20014
(signatories: District of Columbia, Maryland, Pennsylvania, Virginia, West Virginia)

Interstate Sanitation Commission
10 Columbus Circle
New York, N.Y. 10019
(states: Connecticut, New Jersey, New York)

Klamath River Compact Commission
280 Main Street
Klamath Falls, Oreg. 97601
(mailing address: P.O. Box 388, Sacramento Calif. 95802)
(states: California, Oregon)

New England Interstate Water Pollution Control Commission
607 Boylston Street
Boston, Mass. 02116
(states: Connecticut, Maine, Massachusetts, New Hampshire, New York, Rhode Island, Vermont)

Ohio River Valley Water Sanitation Commission (ORSANCO)
414 Walnut Street
Cincinnati, Ohio 45202
(states: Illinois, Indiana, Kentucky, New York, Ohio, Pennsylvania, Virginia, West Virginia)

C. STATE AND TERRITORY

ALABAMA

Alabama Water Improvement Commission
State Office Building
Montgomery, Ala. 36104

ALASKA

Water Pollution Control Section
Water and Air Quality Control Division
Department of Environmental Conservation
Pouch O
Juneau, Alaska 99881

ARIZONA

Bureau of Water Quality Control
Arizona Department of Health Services
1740 West Adams Street
Phoenix, Ariz. 85007

ARKANSAS

Pollution Control Commission
1100 Harrington Avenue
Little Rock, Ark. 72201

CALIFORNIA

State Water Resources Control Board
1416 Ninth Street
Sacramento, Calif. 95814

COLORADO

Water Quality Control Division
4210 East 11th Avenue
Denver, Colo. 80220

CONNECTICUT

Department of Environmental Protection
165 Capitol Avenue
Hartford, Conn. 06115

DELAWARE

Division of Environmental Control
Department of Natural Resources and Environmental Control
Dover, Del. 19901

DISTRICT OF COLUMBIA

Water Quality Division
Bureau of Air and Water Quality Control
614 H Street, N.W.
Washington, D.C. 20001

FLORIDA

Department of Environmental Regulation
2562 Executive Center Circle, E
Montgomery Building
Tallahassee, Fla. 32301

GEORGIA

Water Quality Environmental Protection Division
Department of Natural Resources
244 Washington Street, S.W.
Atlanta, Ga. 30334

GUAM

Water Pollution Control Center
P.O. Box 2999
Agana, Guam 96910

HAWAII

Department of Health
P.O. Box 3378
Honolulu, Hawaii 96801

IDAHO

Assistant Administrator and Director of Environmental Services
Statehouse
Boise, Idaho 83720

ILLINOIS

Pollution Control Board
309 West Washington Street
Suite 300
Chicago, Ill. 60606

INDIANA

Stream Pollution Control Board
1330 West Michigan Street
Indianapolis, Ind. 46202

IOWA

Water Quality Management Division
Department of Environmental Quality
P.O. Box 3326
Des Moines, Iowa 50316

KANSAS

Bureau of Water Quality
Division of Environment
Building 740, Forbes Air Force Base
Topeka, Kans. 66620

KENTUCKY

Division of Sanitary Engineering
Department for Natural Resources and Environmental Protection
275 East Main Street
Frankfort, Ky. 40601

LOUISIANA

Stream Control Commission
P.O. Drawer FC
University Station
Baton Rouge, La. 70803

MAINE

Environmental Improvement Commission
Statehouse
Augusta, Maine 04330

MARYLAND

Division of Water and Sewerage
201 West Preston Street
Baltimore, Md. 21201

MASSACHUSETTS

Division of Water Resources
100 Cambridge Street
Boston, Mass. 02202

MICHIGAN

Environmental Protection Branch
Department of Natural Resources
Stevens T. Mason Building
Lansing, Mich. 48926

MINNESOTA

Chief
Water Supply and General Engineering
Department of Health
717 Delaware Street, S.E.
Minneapolis, Minn. 55440

MISSISSIPPI

Air and Water Pollution Control
P.O. Box 827
Jackson, Miss. 39205

MISSOURI

Division of Environmental Quality
Department of Natural Resources
P.O. Box 1368
Jefferson, City, Mo. 65101

MONTANA

Water Quality Bureau
Department of Health and Environmental Sciences
Helena, Mont. 59601

NEBRASKA

Water Pollution Control Division
Department of Environmental Control
P.O. Box 94653
Lincoln, Nebr. 68509

NEVADA

Environmental Protection Services
201 South Fall Street
Carson City, Nev. 89701

NEW HAMPSHIRE

Water Supply and Pollution Control Commission
105 Loudon Road
P.O. Box 95
Concord, N.H. 03301

NEW JERSEY

Water Pollution Control Monitoring Surveillance and Enforcement
1474 Prospect Street
P.O. Box 2809
Trenton, N.J. 08625

NEW MEXICO

Water Quality Division
Environmental Improvement Agency
P.O. Box 2348
Santa Fe, N. Mex. 87501

NEW YORK

Division of Pure Waters
Department of Environmental Conservation
50 Wolf Road
Albany, N.Y. 12233

NORTH CAROLINA

Division of Environmental Management
Department of Natural and Economic Resources
P.O. Box 27687
Raleigh, N.C. 27611

NORTH DAKOTA

Division of Water Supply and Pollution Control
State Department of Health
State Capitol Building
Bismarck, N. Dak. 58501

OHIO

State of Ohio Environmental Protection Agency
P.O. Box 1049
Columbus, Ohio 43216

OKLAHOMA

Environmental Health Services
Northeast Tenth and Stonewall
Oklahoma City, Okla. 73105

OREGON

Department of Environmental Quality
1234 Morrison Street, Southwest
Portland, Oreg. 97201

PENNSYLVANIA

Department of Environmental Resources
P.O. Box 1467
Harrisburg, Pa. 17120

PUERTO RICO

Department of Health
Water Pollution Control
Stop 19
Santurce, P.R. 00908

RHODE ISLAND

Principal Sanitary Engineer
Health Building
Davis Street
Providence, R.I. 02908

SOUTH CAROLINA

Bureau of Wastewater and Stream Quality Control
Department of Health and Environmental Control
J. Marion Sims Building
Columbia, S.C. 29201

SOUTH DAKOTA

Department of Environmental Protection
Capitol Building
Pierre, S. Dak. 57501

TENNESSEE

Division of Water Quality Control
621 Cordell Hull Building
Nashville, Tenn. 37219

TEXAS

Water Quality Board
P.O. Box 13246, Capitol Station
Austin, Tex. 78701

UTAH

Water Pollution Committee
44 Medical Drive
Salt Lake City, Utah 84113

VERMONT

Commissioner
Water Resources
Montpelier, Vt. 05602

VIRGINIA

State Water Control Board
P.O. Box 1143
Richmond, Va. 23230

VIRGIN ISLANDS

Air and Water Pollution Control
Charlotte Amalie
St. Thomas, V.I. 00801

WASHINGTON

Department of Ecology
P.O. Box 829
Olympia, Wash. 98501

WEST VIRGINIA

Division of Water Resources
Department of Natural Resources
1201 Greenbriar Street, E.
Charleston, W. Va. 25311

WISCONSIN

Department of Natural Resources
Box 450
Madison, Wis. 53701

WYOMING

Water Quality Division
Department of Environmental Quality
State Office Building
Cheyenne, Wyo. 82002

Chapter 26
PUBLISHERS

Academic Press
111 Fifth Avenue
New York, N.Y. 10003

Addison-Wesley Publishing Co.
Jacob Way
Reading, Mass. 01867

American Association for the Advancement of Science
1515 Massachusetts Avenue, N.W.
Washington, D.C. 20005

American Chemical Society
1155 Sixteenth Street, N.W.
Washington, D.C. 20036

American Elsevier Publishing Co.
52 Vanderbilt Avenue
New York, N.Y. 10017

American Institute of Chemical Engineers
36 East Forty-seventh Street
New York, N.Y. 10017

American Society for Testing and Materials
1916 Race Street
Philadelphia, Pa. 19103

American Society of Civil Engineers
345 East Forty-seventh Street
New York, N.Y. 10017

American Water Resources Association
Box 434
Urbana, Ill. 61801

American Water Works Association
Publication Sales Department
2 Park Avenue
New York, N.Y. 10016

Ann Arbor Science Publishers
P.O. Box 1425
Ann Arbor, Mich. 48106

Edward Arnold
25 Hill Street
London, Engl. WIX 8LL

Barnes & Noble
105 Fifth Avenue
New York, N.Y. 10003

Blackwell Scientific Publishing
Osney, Mead
Oxford, Engl.

R.R. Bowker Co.
1180 Avenue of the Americas
New York, N.Y. 10036

Butterworth & Co.
88 Kingsway
London, W.C. 2, Engl.

Publishers

Cahners Publishing Co.
221 Columbus Avenue
Boston, Mass. 02116

Cambridge University Press
32 East Fifty-seventh Street
New York, N.Y. 10022

Chemical Rubber Co.
18901 Cranwood Parkway
Cleveland, Ohio 44128

Clearinghouse for Federal Scientific and Technical Information
5285 Port Royal Road
Springfield, Va. 22151

Colorado State University
Engineering Research Center
Fort Collins, Colo. 80521

Columbia University Press
440 West 110th Street
New York, N.Y. 10025

Cornell University Press
124 Roberts Place
Ithaca, N.Y. 14850

Walter de Gruyter
162 Fifth Avenue
New York, N.Y. 10010

Marcel Dekker
95 Madison Avenue
New York, N.Y. 10016

Doubleday & Co.
501 Franklin Avenue
Garden City, N.Y. 11530

Elsevier/North Holland
335 Jan Van Galenstraat
P.O. Box 211
Amsterdam-W, Netherlands

Environment Information Center
Film References Department
124 East Thirty-ninth Street
New York, N.Y. 10016

W.H. Freeman and Co.
660 Market Street
San Francisco, Calif. 94104

Gale Research Co.
Book Tower
Detroit, Mich. 48226

Grossman Publishers
44 West Fifty-sixth Street
New York, N.Y. 10019

Harper and Row Publishers
49 East Thirty-third Street
New York, N.Y. 10016

Harvard University Press
79 Garden Street
Cambridge, Mass. 02138

Her Imperial Majesty's Stationery Office
Department of Health and Social Security, Welsh Office
Ministry of Housing and Local Government
London, Engl.

Holt, Rinehart and Winston
383 Madison Avenue
New York, N.Y. 10017

Houghton Mifflin Co.
2 Park Street
Boston, Mass. 02107

IFI/Plenum Corporation
227 West Seventeenth Street
New York, N.Y. 10011

Indiana University
Water Resources Research Center
Bloomington, Ind. 47401

Information Retrieval
1911 Jefferson Davis Highway
Suite 907
Arlington, Va. 22203

Institute for Scientific Information
Publications Division
325 Chestnut Street
Philadelphia, Pa. 19106

International Atomic Energy Agency
UNIPUB, Inc.
P.O. Box 433
New York, N.Y. 10016

Interscience Publishers
Division of John Wiley & Sons
605 Third Avenue
New York, N.Y. 10016

Iowa State University Press
Ames, Iowa 50010

John Hopkins University Press
Baltimore, Md. 21218

La Motte Chemical Products Co.
Chestertown, Md. 21620

J.B. Lippincott Co.
East Washington Square
Philadelphia, Pa. 19105

McGraw-Hill Book Co.
1221 Avenue of the Americas
New York, N.Y. 10020

Macmillan Company
886 Third Avenue
New York, N.Y. 10022

MIT Press
28 Carleton Street
Cambridge, Mass. 02142

National Academy of Sciences
2101 Constitution Avenue
Washington, D.C. 20418

National Technical Information Service (NTIS)
Department of Commerce
Springfield, Va. 22151

North Holland Publishers
Amsterdam, Holland

Noyes Data Corporation
118 Mill Road
Park Ridge, N.J. 07656

Ocean Engineering Information Service
P.O. Box 989
La Jolla, Calif. 92037

Organization for Economic Cooperation and Development
2 Rue Andre-Pascal
75 Paris 16e, France

Pergamon Press
Maxwell House
Fairview Park, Elmsford, N.Y. 10523

Plenum Publishing Corporation
227 West Seventeenth Street
New York, N.Y. 10011

Prentice-Hall
Englewood Cliffs, N.J. 07632

Princeton University Press
Princeton, N.J. 08540

D. Reidel Publishing Corporation
160 Old Derby Street
Hingham, Mass. 02043

Publishers

Reinhold Publishing Corporation
600 Summer Street
Stamford, Conn. 06940

Resources for the Future
1755 Massachusetts Avenue, N.W.
Washington, D.C. 20036

W.B. Saunders Company
218 West Washington Square
Philadelphia, Pa. 19105

Sierra Club Books
530 Bush Street
San Francisco, Calif. 94108

Springer-Verlag New York
175 Fifth Avenue
New York, N.Y. 10010

Charles C Thomas, Publisher
301-327 East Lawrence Avenue
Springfield, Ill. 62717

UNESCO
National Agency for International Publications
317 East Thirty-fourth Street
New York, N.Y. 10016

UNESCO
Unipub
P.O. Box 433
New York, N.Y. 10016

United Nations Publications
United Nations, Room LX 2300
New York, N.Y. 10017

U.S. Department of Commerce Clearinghouse
Springfield, Va. 22151

U.S. Department of the Interior
Geological Survey
Washington, D.C. 20242

U.S. Environmental Protection Agency
4676 Columbia Parkway
Cincinnati, Ohio 45206

U.S. Government Printing Office
Washington, D.C. 20402

U.S. Public Health Service
Office of Public Information
Ballston Center Tower 2
801 North Randolph
Arlington, Va. 22203

University Microfilms International
300 North Zeeb Road
Ann Arbor, Mich. 48106

University of Arizona Press
P.O. Box 3398
College Station
Tucson, Ariz. 85700

University of California
Marine Science Institute
Oil Spill Information Center
Santa Barbara, Calif. 93106

University of California Press
2223 Fulton Street
Berkeley, Calif. 94720

University of Chicago Press
5801 Ellis Avenue
Chicago, Ill. 60637

University of Illinois
Department of Civil Engineering
Engineering Publications Office
112 Engineering Hall
Urbana, Ill. 61801

University of Texas
Center for Research in Water Resources
Austin, Tex. 78757

University of Washington Press
Seattle, Wash. 98105

University of Wisconsin
Water Resources Center
Hydraulic and Sanitary Laboratory
Madison, Wis. 53706

Van Nostrand Reinhold Company
450 West Thirty-third Street
New York, N.Y. 10001

Water Information Center
Water Research Building
Manhasset Isle
Port Washington, N.Y. 11050

Water Pollution Control Federation
3900 Wisconsin Avenue
Washington, D.C. 20016

John Wiley and Sons
605 Third Avenue
New York, N.Y. 10016

H.W. Wilson Company
950 University Avenue
Bronx, N.Y. 10452

Part IV

APPENDIXES

Appendix A
GLOSSARY

The following brief glossary has been developed over a ten-year period of teaching a water pollution course. The terms included are commonly encountered in reading water pollution literature. The glossary was developed by adapting terms from Ruttner (1965), p. 65, and Mackenthun and Ingram (1967), p. 79. If more extensive treatment is desired, see chapter 8, "Lexicons," p. 41, for an in-depth treatment of dictionaries, glossaries, and source books related to water pollution.

Aerobic (Aerobe): Organisms that can prosper only when oxygen is present more or less abundantly. (Gr. aer air, bios life)

Alga (plural Algae): The simplest of all plant forms, having neither roots, stems, nor leaves. Algae range in size from microscopic single cells to branching forms one hundred feet or more in length.

Algicide: A specific chemical highly toxic to algae. Algicides are often applied to water to control nuisance algal blooms.

Algology: The study of algae.

Amphibious Organism: An organism adapted for life on land or in water.

Anadromous Fishes: Fishes that spend a part of their life in the sea or lakes, but ascend rivers at more or less regular intervals to spawn. Examples are sturgeon, shad, salmon, trout, and striped bass.

Anaerobic (Anaerobe): Organisms which either by obligation or facultatively thrive in the absence of oxygen. (Gr. an without, aer air, bios life)

Annelids: Segmented worms, as distinguished from the nonsegmented roundworms and flatworms. Most are marine, however, many live in soil or fresh water. Aquatic forms may establish dense populations in the presence of rich organic deposits. Common examples of segmented worms are earthworms, sludgeworms, sandworms, and leeches.

Glossary

Aquatic: Living in water.

Autotrophic: The nutrition of those plants that are able to construct organic matter from inorganic. Denotes those green plants and forms of bacteria that do not require organic carbon or nitrogen, but can form their own food out of inorganic salts and carbon dioxide. (Gr. autos self, trophein to nourish)

Bacteria: Simple, colorless, one-celled plants, most of which are unable to manufacture their own food using sunlight. Although some bacteria cause diseases, others fill an indispensable ecological role as decomposers.

Benthal: The region of the shore and the bottom of waters. (Gr. benthos depth)

Benthic Region: The bottom of all waters; the substratum that supports the benthos.

Benthos: Bottom-dwelling organisms, the benthos comprise: (1) sessile animals such as the sponges, barnacles, mussels, and oysters, some of the worms, and many attached algae; (2) creeping forms, such as snails and flatworms; and (3) burrowing forms which include most clams and worms.

Bioassay: A determination of the concentration of a given material by comparison with a standard preparation, or the determination of the quantity necessary to affect a test animal under stated laboratory conditions.

Biota: All living organisms of a region.

Bloom: A readily visible concentrated growth or aggregation of plankton (plant and animal).

Blue-Green Algae: A group of algae with a blue pigment in addition to the green chlorophyll. A stench is often associated with the decomposition of dense blooms of blue-green algae in fertile lakes.

Carnivore: An animal that lives by eating the flesh of other animals. See also Herbivore, Omnivore.

Catadromous Fishes: Fishes that feed and grow in fresh water, but return to the sea to spawn. The best-known example is the American eel.

Chlorophyll: The green pigments of plants that produce the green color essential to photosynthesis. (Gr. chloros green, phyllon leaf)

Coarse or Rough Fish: Those species of fish considered to be of poor fighting quality when taken on tackle and of poor food quality. These fish may be undesirable in a given situation, but at times may be classified differently depending upon their usefulness. Examples include carp, goldfish, gar, sucker, bowfin, gizzard shad, goldeneye, mooneye, and certain kinds of catfish.

Crustacea: Mostly aquatic animals with rigid outer coverings, jointed appendages, and gills. Examples are crayfish, crabs, barnacles, water fleas, and sow bugs.

Cubic Feet Per Second - (cfs): A measure of discharge. The amount of water passing a given point, expressed as number of cubic feet in each second.

Current: The constant downstream flow of water in a river or stream resulting from water's tendency to move downslope toward sea level.

Decomposers: Living plants and animals, but chiefly fungi and bacteria, that live by extracting energy from the decaying tissues of dead plants and animals. In the process they release simple chemical compounds stored in the dead bodies and make them available for use by green plants.

Denitrification: Reduction from nitrate to nitrite and further to elemental nitrogen. (Lat. de from, nitrum saltpeter, facere to make)

Diatom: A single-celled alga encased in an intricately etched silica shell formed of two halves that fit together like the lid on a box. Diatoma are important primary producers in rivers, streams, and other bodies of water.

Discharge: Outflow. The flow of a stream, canal, or aquifer. It is also common to refer to the discharge of a canal or stream into a lake, river, or an ocean.

Domestic Use: Use of water in homes and on lawns, including use for laundry, washing cars, cooling, and swimming pools.

Ecology: The study of the relationships of organisms to their environment. (Gr. oikos house, logos discourse)

Ecosystem: The functioning together of the biological community and the non-living environment.

Emergent Aquatic Plants: Plants that are rooted at the bottom but project above the water surface. Examples are cattails and bulrushes.

Environment: All the external conditions surrounding a living thing.

Epilimnion: The turbulent superficial layer of a lake lying above the thermocline which does not have a permanent thermal stratification. (Gr. epi on, limne lake)

Erosion: The wearing away of areas of the earth's surface by water, wind, ice, and other natural forces.

Estuary: Commonly an arm of the sea at the lower end of a river. Estuaries are often enclosed by land except at channel entrance points.

Eutrophic: Waters with a good supply of nutrients and hence a rich organic production. (Gr. eu well, trophein to nourish)

Eutrophication: The intentional or unintentional enrichment of water.

Facultative Aerobe: An organism that although fundamentally an anaerobe can grow in the presence of free oxygen.

Facultative Anaerobe: An organism that although fundamentally an aerobe can grow in the absence of free oxygen.

Fauna: The entire animal life of a region.

Floating Aquatic Plants: Plants that wholly, or in part, float on the surface of the water. Examples are water lilies, water shields, and duckweeds.

Flora: The entire plant life of a region.

Food Chain: A series of plants and animals linked by their food relationships. The passage of energy and materials from primary producers (green plants) through a succession of consumers. Green plants, plant-earing insects, and an insect-eating fish would form a simple food chain. See also Food Web.

Food Web: A system of interlocking food chains. Since few animals rely on a single food source and since a given food source rarely is consumed exclusively by a single kind of animal, the separate food chains in any natural community interlock and form a web.

Fungi (singular Fungus): Simple or complex organisms without chlorophyll. The simpler forms are one-celled, the higher forms have branched filaments and complicated life cycles. Examples of fungi are molds, yeasts, and mushrooms.

Green Algae: Algae that have pigments similar in color to those of higher green plants. Common forms produce algal mats or floating "moss" in lakes.

Headwaters: The sources or upper parts of a river.

Herbivore: An organism that feeds on vegetation. An animal that eats plants, thus making the energy stored in plants available to carnivores. See also Carnivore, Omnivore.

Heterotrophic: The nutrition of plants and animals that are dependent on organic matter for food. (Gr. _heteros_ other, _trophein_ to nourish)

Heterotrophic Organisms: Organisms that are dependent on organic matter for food.

Higher Aquatic Plants: Flowering aquatic plants. (These are separately categorized herein as Emergent, Floating, and Submerged Aquatic Plants.)

Hydrobiology: The study of life in water. (Gr. _hydor_ water, _bios_ life, _logos,_ discourse)

Hypolimnion: The deep layer of a lake lying below the thermocline and removed from surface influences. (Gr. _hypo_ under, _limne_ lake)

Ichthyologist: A specialist in the study of fishes.

Insecticide: Substances or a mixture of substances intended to prevent, destroy, or repel insects.

Invertebrates: Animals without backbones.

Larva (plural Larvae): An active immature stage in an animal's life history during which its form differs from that of the adult. The caterpillar, for example, is the larval stage in the life history of a butterfly; the tadpole is the larva of a frog. See also Nymph.

LD50: See Median Lethal Dose.

Lenitic or Lentic Environment: Standing water and its various intergrades. Examples of lenitic environments are lakes, ponds, and swamps.

Life Cycle: The series of stages in the form and mode of life of an organism, i.e., the stages between successive recurrences of a certain primary stage such as the spore, fertilized egg, seed, or resting cell.

Limnetic Zone: The openwater region of a lake. This region supports plankton and fish as the principal plants and animals.

Limnology: The study of the physical, chemical, and biological aspects of inland waters.

Littoral: The shoreward region of a body of water. (Lat. _littus,_ prop. _litus_ shore)

Lotic Environment: Running waters such as streams or rivers.

Macroorganisms: Plant, animal, or fungal organisms visible to the unaided eye.

Macrophytes: Large plants. (Gr. _makros_ great, _phyton_ plant)

Median Lethal Dose (LD50): The dose lethal to 50 percent of a group of test organisms for a specified period. The dose material may be ingested or injected.

Median Tolerance Limit (TL_m): The concentration of the tested material in a suitable diluent (experimental water) at which just 50 percent of the test animals are able to survive for a specified period of exposure.

Metalimnion: See Thermocline.

Microorganism: Any minute organism invisible or barely visible to the unaided eye.

Microphytes: Small plants. (Gr. _mikros_ small, _phyton_ plant)

Glossary

Motile: Exhibiting or capable of spontaneous movement.

Mycology: The study of fungi.

Nekton: The powerful swimmers among the plankton animals that to a large degree are capable of moving about voluntarily. (Gr. nektos swimming)

Nematoda: Unsegmented roundworms or threadworms. Some are free living in soil, freshwater, and salt water. Some are found living in plant tissue, others live in animal tissue as parasites.

Neuston: Organisms resting or swimming on the surface film of the water; the community of the surface film of the water. (Gr. nein to swim)

Nymph: The immature, preadult form of an insect, such as a dragonfly, whose life history is characterized by incomplete metamorphosis. Nymphs after hatching live and feed in water until they molt and become adults. See also Larva.

Oligothermal: Confined to a range of low temperatures. (Gr. oligos small, therme heat)

Oligotrophic: Waters with a small supply of nutrients and hence a small organic production. (Gr. oligos small, trophein to nourish)

Omnivore: An animal that habitually eats both plants and animals. See also Carnivore, Herbivore.

Organic: Pertaining to anything that is or ever was alive or produced by a living plant or animal.

Organic Detritus: The particulate remains of disintegrated plants and animals.

Parasite: An organism that lives on or in a host organism from which it obtains nourishment at the expense of the latter during all or part of its existence.

Periphyton: The association of aquatic organisms attached or clinging to stems and leaves of rooted plants or other surfaces projecting above the bottom.

pH: The negative logarithm of the hydrogen ion concentration expressed in gram equivalents.

Photosynthesis: Construction of organic matter (carbohydrate) from CO_2 and H_2O with the aid of the energy of light. (Gr. phos, photos light, synthesis placing together)

Phytoplankton: The plant portion of the plankton. (Gr. phyton plant)

Piscicide: Substances or a mixture of substances intended to destroy or control fish populations.

Plankton: The community of the free water. (Gr. n. of planktos wandering)

Pollution: The fouling of water or air with sewage, heat, industrial wastes, or other contaminants, making them unfit to support many forms of life.

Primary Producers: Green plants, the basic link in any food chain. By means of photosynthesis, green plants manufacture the food on which all other living things ultimately depend.

Producers: Organisms that synthesize their own organic substance from inorganic substances, for example, plants.

Production (Productivity): A time-rate unit of the total amount of organisms grown.

Productivity: The total weight or number of living things produced in a functioning natural community. Productivity depends on the interaction of the life and the environment.

Protozoa: Organisms consisting either of a single cell or of aggregates of cells, each of which performs all the essential functions in life. They are mostly microscopic in size and largely aquatic.

Protozoologist: A specialist in the study of protozoa.

Saprobic: Organisms living in water polluted with organic materials. (Gr. <u>sapros</u> rotten, bios life)

Scavenger: An animal that eats the dead remains and wastes of other animals and plants.

Sediment: Fragmental mineral material transported or deposited by water or air.

Sphaerotilus: A slime-producing, nonmotile, sheathed, filamentous, attached bacterium. Great masses are often broken from their "holdfasts" by currents and are carried floating downstream in gelatinous flocks.

Stenotopic: An organism with a narrow range of tolerance for a particular environmental factor, i.e., stenothermal in relation to temperature. (Gr. <u>stenos</u> narrow, <u>topos</u> place)

Stenotopic Organisms: Organisms with a narrow range of tolerance for a particular environmental factor. Examples are trout, stonefly nymphs, and so forth.

Submerged Aquatic Plant: A plant that is continuously submerged beneath the surface of the water. Examples are the pondweed and coontail.

Symbiosis: Two organisms of different species living together, one or both of which may benefit and neither is harmed.

Thermal Gradient: The temperature difference within a layer 1 m. thick.

Thermocline: The layer of water in a lake between the epilimnion and hypolimnion in which the temperature exhibits the greatest difference in a vertical direction. (Gr. therme heat, klinein to slope)

Tubificidae: Aquatic segmented worms that exhibit marked population increases in aquatic environments containing organic decomposable wastes.

Vertebrate: Animals with backbone.

Warm and Coldwater Fish: Warmwater fish include black bass, sunfish, catfish, gar, and others, whereas coldwater fish include salmon and trout, whitefish, miller's thumb, and blackfish. The temperature factor determining distribution is set by adaptation of the eggs to warm or cold water.

White Water: Swiftly flowing, frothy water in a rapids.

Zooplankton: The animal portion of the plankton. (Gr. zoion animal)

Appendix B
SELECTED READINGS

The area covered by the term "water pollution" is so broad one is frequently requested to suggest a basic text for a good introduction and basic understanding of the principles of water pollution. A single adequate reference of this nature is not available. Instead of a single reference, a selected reading list is provided to acquaint the uninformed reader with a general overview of water pollution. The references were selected to provide a balanced introduction of the entire water pollution field. The suggested readings are not intended to provide a complete coverage of all pollutants nor extensive treatment of those included, but to provide a selected reading list for entering the area of water pollution. Having completed the readings in this section, the reader will possess an adequate background to consult the appropriate sections for more in depth readings, or enter into an intensive search of available literature using the reference sources provided on specific topics.

A. HISTORY OF WATER POLLUTION

Warren, C[harles]. E. BIOLOGY AND WATER POLLUTION CONTROL. Philadelphia: W.B. Saunders Co., 1971. Chapters 1, 2, 4, pp. 3-14, 24-39.

B. STATE-OF-THE-ART

Commoner, B.; Corr, M.; and Stamler, P.F. "The Causes of Pollution." ENVIRONMENT 13 (April 1971): 2-19.

Ridker, R.G. "Population and Pollution in the United States." SCIENCE 176 (9 June 1972): 1085-90.

Wilber, C[harles]. G. THE BIOLOGICAL ASPECTS OF WATER POLLUTION. Springfield, Ill.: Charles C Thomas Publishers, 1969. Chapter 2, pp. 17-30.

C. AQUATIC ECOLOGY

Hynes, H.B.N. THE ECOLOGY OF RUNNING WATERS. Toronto: University of Toronto Press, 1970. Chapters 1-7, pp. 1-112.

Odum, E.P. FUNDAMENTALS OF ECOLOGY. 3d ed. Philadelphia: W.B. Saunders Company, 1971. Chapters 11-13, pp. 295-362.

Whitton, B.A., ed. RIVER ECOLOGY. Studies in Ecology, vol. 2. Berkeley and Los Angeles: University of California Press, 1975. Chapter 2, 15, pp. 39-80, 375-402.

D. TYPES OF POLLUTION (SELECTED EXAMPLES)

Hunt, G.S. "The Direct Effects on Some Plants and Animals of Pollution in the Great Lakes." BIOSCIENCE (March 1965): 181-86.

Warren, C[harles]. E. BIOLOGY AND WATER POLLUTION CONTROL. Philadelphia: W.B. Saunders Company, 1971. Chapter 6, pp. 57-63.

E. TOXIC MATERIALS

Ahmed, A.K. "PCB's in the Environment." ENVIRONMENT 18 (March 1976): 6-11.

Rudd, R.L. "Pesticides." In ENVIRONMENT, edited by William W. Murdoch, pp. 325-53. Sunderland, Mass.: Sinauer Associates, 1975.

Selikoff, I.S., ed. "Hazards of Mercury." ENVIRONMENTAL RESEARCH 4 (March 1971): 1-69.

Wood, J.M. "Biological Cycles for Toxic Elements in the Environment." SCIENCE 183 (15 March 1974): 1049-52.

Woodwell, G.M.; Craig, P.P.; and Johnson, H.A. "DDT in the Biosphere: Where Does it Go?" SCIENCE 174 (10 December 1971): 1101-18.

F. OIL POLLUTION

Blumer, M[ax]. "Scientific Aspects of the Oil Spill Problem." ENVIRONMENTAL AFFAIRS 1 (April 1971): 54-73.

Mitchell, R[alph]., ed. WATER POLLUTION MICROBIOLOGY. New York: Wiley-Interscience, 1972. Chapter 6, pp. 119-51.

G. THERMAL POLLUTION

Becker, C.D. "Columbia River Thermal Effects Study: Reactor Effluent Problems." JOURNAL WATER POLLUTION CONTROL FEDERATION 45 (May 1973): 850-68.

Brown, G[eorge].W., and Krygier, J.T. "Effects of Clear-Cutting on Stream Temperature." WATER RESOURCES RESEARCH 6 (August 1970): 1133-39.

Bush, R.M.; Welch, E.B.; and Mar, B.W. "Potential Effects of Thermal Discharges on Aquatic Systems." ENVIRONMENTAL SCIENCE AND TECHNOLOGY 8 (June 1974): 561-68.

Cairns, J[ohn]., Jr.; Heath, A.G.; and Parker, B.C. "Temperature Influence on Chemical Toxicity to Aquatic Organisms." JOURNAL WATER POLLUTION CONTROL FEDERATION 47 (February 1975): 267-80.

Lehmkuhl, D.M. "Change in Thermal Regime as a Cause of Reduction of Benthic Fauna Downstream of a Reservoir." JOURNAL CANADA FISHERIES RESEARCH BOARD 29 (September 1972): 1329-32.

H. SUSPENDED MATTER

Barton, B.A. "Short-Term Effects of Highway Construction on the Limnology of a Small Stream in Southern Ontario." FRESHWATER BIOLOGY 7 (April 1977): 99-108.

Brown, G[eorge]. W., and Krygier, J.T. "Clear-Cut Logging and Sediment Production in the Oregon Coast Range." WATER RESOURCES RESEARCH 7 (October 1971): 1189-98.

Chutter, F.M. "The Effects of Silt and Sand on the Invertebrate Fauna of Streams and Rivers." HYDROBIOLOGIA 34 (1969): 57-76.

I. RADIOACTIVITY

Maxwell, K.E. CHEMICALS AND LIFE. Belmont, Calif.: Dickenson Publishing Company, 1970. Chapter 11, pp. 333-50.

Wilber, C[harles]. G. THE BIOLOGICAL ASPECTS OF WATER POLLUTION. Springfield, Ill.: Charles C Thomas Publishers, 1969. Chapter 11, pp. 195-223.

Selected Readings

J. NUISANCE ORGANISMS

Mackenthun, K[enneth].M. THE PRACTICE OF WATER POLLUTION BIOLOGY. Washington, D.C.: U.S. Department of Interior, Federal Water Pollution Control Administration, 1969. Chapter 15, pp. 221-36.

K. PATHOGENS

Gerba, C.P.; Wallis, C.; and Melnick, J.L. "Viruses in Water: the Problem, Some Solutions." ENVIRONMENTAL SCIENCE AND TECHNOLOGY 9 (December 1975): 1122-26.

Mitchell, R[alph]., ed. WATER POLLUTION MICROBIOLOGY. New York: Wiley-Interscience, 1972. Chapters 9, 16, pp. 207-41, 365-88.

Taylor, F.B. "Viruses--What Is Their Significance in Water Supplies?" AMERICAN WATER WORKS ASSOCIATION JOURNAL 66 (May 1974): 306-11.

L. EUTROPHICATION

Hutchinson, G.E. "Eutrophication." AMERICAN SCIENTIST 61 (May-June 1973): 269-79.

Likens, G.E., ed. "Nutrients and Eutrophication: The Limiting-Nutrient Controversy." In LIMNOLOGY AND OCEANOGRAPHY, edited by Y.H. Edmondson, pp. 3-13. Vol. 1. Special Symposia, 1972.

Schindler, D. "Eutrophication and Recovery in Experimental Lakes: Implications for Lake Management." SCIENCE 184 (24 May 1974): 897-99.

M. ORGANIC POLLUTION

Bartsch, A.F., and Ingram, W.M. "Stream Life and the Pollution Environment." PUBLIC WORKS 90, no. 7 (1959): 104-10.

N. WASTE TREATMENT

Mitchell, R[alph]., ed. WATER POLLUTION MICROBIOLOGY. New York: Wiley-Interscience, 1972. Chapter 16, pp. 365-88.

Warren, C[harles]. E. BIOLOGY AND WATER POLLUTION CONTROL. Philadelphia: W.B. Saunders Co., 1971. Chapter 21, pp. 349-72.

O. WATER RENOVATION AND REUSE

Allen, J. "Sewage Farming." ENVIRONMENT 15 (March 1973): 36-41.

Dworkin, D.M. "Water Reuse: A Flexible and Efficient Management Alternative for Municipal Water Supply." WATER RESOURCES BULLETIN 11 (October 1975): 607-15.

English, J.N., and Mitchell, T.M. "Water Reclamation and Reuse." JOURNAL WATER POLLUTION CONTROL FEDERATION 48 (June 1976): 1174-80.

P. POLITICAL AND LEGAL ASPECTS

Blazey, D.R. "Legal Perspectives on Water Quality Management." A.S.C.E. JOURNAL HYDRAULICS DIVISION 101 (May 1975): 595-601.

Warren, C.E. BIOLOGY AND WATER POLLUTION CONTROL. Philadelphia: W.B. Saunders Co., 1971. Chapter 3, pp. 15-23.

Q. FUTURE PROBLEMS AND PROPOSALS

Harvey, H.H. "Aquatic Environmental Quality: Problems and Proposals." JOURNAL CANADA FISHERIES RESEARCH BOARD 33, no. 11 (November 1976): 2634-70.

Warren, C[harles]. E. BIOLOGY AND WATER POLLUTION CONTROL. Philadelphia: W.B. Saunders Co., 1971. Chapter 23, pp. 387-95.

INDEXES

AUTHOR INDEX

In addition to authors, this index includes all editors, compilers, translators, and other contributors to works cited in the text. References are to page numbers and alphabetization is letter by letter.

Author Index

C

D

Author Index

G

H

I

J

K

L

M

N

O

P

Q

R

S

T

U

V

W

Y

Z

TITLE INDEX

This index includes all titles of books, reports, and proceedings cited in the text. References are to page number and alphabetization is letter by letter.

A

Title Index

D

E

F

G

H

I

L

M

N

Q

R

S

T

U

V

W

Y

Z

SUBJECT INDEX

This index is alphabetized letter by le-ter and references are to page numbers.

N

O

P

R

S